APR 2 9 2003

D0428817

PROPERTY OF:
RANCHO MIRAGE PUBLIC LIBRARY
42-520 Bob Hope Drive
Rancho Mirage, CA 92270
(760) 341-READ (7323)

UNLOCKING
THE SKY

BOOKS BY SETH SHULMAN

Owning the Future

*The Threat at Home: Confronting the
Toxic Legacy of the U.S. Military*

UNLOCKING THE SKY

GLEN HAMMOND CURTISS
AND THE RACE
TO INVENT
THE AIRPLANE

SETH SHULMAN

HarperCollins*Publishers*

UNLOCKING THE SKY. Copyright © 2002 by Seth Shulman. All rights reserved. Printed in the United States of America. No part of this book may be used or reproduced in any manner whatsoever without written permission except in the case of brief quotations embodied in critical articles and reviews. For information, address HarperCollins Publishers Inc., 10 East 53rd Street, New York, NY 10022.

HarperCollins books may be purchased for educational, business, or sales promotional use. For information, please write: Special Markets Department, Harper-Collins Publishers Inc., 10 East 53rd Street, New York, NY 10022.

Photographs courtesy of the Glenn Curtiss Museum, Hammondsport, N.Y. Permission to reproduce images has been granted by the Archives Division of the National Air and Space Museum, Smithsonian Institution.

FIRST EDITION
Designed by Joseph Rutt
Printed on acid-free paper

Library of Congress Cataloging-in-Publication Data

Shulman, Seth.
    Unlocking the sky : Glenn Curtiss & the race to invent the airplane / Seth Shulman.—1st ed.
        p. cm.
    ISBN 0-06-019633-5
    1. Curtiss, Glenn Hammond, 1878–1930. 2. Airplanes—History. 3. Air pilots—United States—Biography. I. Title.

TL540.C9 S85 2002
629.13'0092—dc21
[B]
                                                2002023882

02 03 04 05 06 ❖/RRD 10 9 8 7 6 5 4 3 2 1

For Benjamin,
who loves inventing

CONTENTS

PROLOGUE   **LANGLEY'S FOLLY**   1

PART I   **REWRITING AVIATION HISTORY**

ONE   **INTRIGUE AT HAMMONDSPORT**   25

TWO   **WRIGHTS AND WRONGS**   41

THREE   **AMERICA OR BUST**   60

PART II   **REACHING FOR THE SKY**

FOUR   **CAPTAINS OF THE AIR**   81

FIVE   **SKY DANCING**   103

SIX   **FLIGHT OF THE *JUNE BUG***   122

SEVEN   SKY KING   144

PART III   WARPED WINGS

EIGHT   GROUNDED   169

NINE   FLIGHT OF A HERO   186

TEN   NEW BEGINNINGS   205

EPILOGUE   ALL BUT THE LEGACY   223

APPENDIX   A PARTIAL LIST OF INVENTIONS BY GLENN CURTISS   231

SOURCES   235

ACKNOWLEDGMENTS   245

INDEX   247

Let us hope that the advent of a successful flying machine, now only dimly foreseen . . . will bring nothing but good into the world; that it shall abridge distance, make all parts of the globe accessible, bring men into closer relation with each other, advance civilization, and hasten the promised era in which there shall be nothing but peace and goodwill among all men.

—OCTAVE CHANUTE, 1894

UNLOCKING
THE SKY

LANGLEY'S FOLLY

*Flight by machines heavier than air is unpractical
and insignificant, if not utterly impossible.*
—SIMON NEWCOMB,
PROMINENT SCIENTIST, 1902

By midafternoon on December 8, 1903, dozens of spectators have gathered on the sunny banks of the Potomac River south of Washington, D.C.

They have come to glimpse the future.

Most have made their way from the city in horse-drawn carriages; some in newfangled motorcars. And now, with overcoats and caps, blankets and field glasses, they huddle against the bitter breeze, chatting excitedly on the riverbank. Scores more have come via the Potomac from nearby wharves, navigating chunks of bobbing ice. They peer from the decks of barges, yachts, and sailboats moored for the unannounced event.

Prominent among the onlookers is a jaunty group of newspaper reporters. Notepads at the ready, they ride the edgy adrenaline of a big story and busy themselves picking out the faces of prominent members of Washington society: Elihu Root, President Teddy Roosevelt's secretary of war, is on hand, as is Army General Wallace F. Randolph and other of the nation's top military brass, scientists, and politicians. Skeptical by trade, the reporters maintain a glib air but, secretly, each recognizes the day's potential. Possibly—just possibly—they could be ringside for the most momentous spectacle of the young century.

This cold, blustery afternoon, all eyes are trained on a large houseboat in the middle of one of the Potomac's widest sections, where it converges with the Anacostia River. Pacing back and forth on the deck of the ungainly craft is the unmistakable figure of Samuel Pierpont Langley, venerable head of the Smithsonian Institution, attired for the occasion in a boating cap and woolen overcoat, his neatly trimmed beard shining white in the afternoon sun.

There on deck, Langley notes how gusty the day has become; the squalls have picked up to 18 miles per hour. From Langley's own account of the day, we know that he is worried as the wind speed rises and the afternoon wears into early evening. But you could never see it in his demeanor. Ever conscious of his role as project director, Langley exudes his usual straight-backed pride and confidence—characteristics sometimes taken for arrogance.

A prominent scientist with an international reputation, Langley is well known by 1903 for his heretical belief that a machine heavier than the air can carry a human being in flight. Yet he has done far more than champion this view. All but abandoning his field of astronomy, Langley has pioneered the embryonic field of "aerial navigation" for fifteen years and published his results widely. Based on the success of Langley's efforts, the U.S. War Department has secretly funded his five-year effort to build a full-scale prototype of the extraordinary new machine he calls an *aerodrome*—Greek for "air runner."

For Langley, today's trial of a full-scale flying machine caps an illustrious career. Now sixty-nine years old, he can boast of successes in both astronomical and aerodynamic research; he has published hundreds of scholarly articles, and he holds honorary degrees from six universities. As director of the Smithsonian Institution for the past seven years, he has advised two U.S. presidents on scientific matters.

Langley's long years of experimentation with more than one hundred aircraft models and his emerging grasp of aeronautical principles have led arduously to this December day in 1903. Yet, despite his imposing manner, his credentials, and his official backing, it strains the credulity of the spectators gathered along the Potomac's banks to believe his machine will actually fly. *No one has ever successfully flown an airplane in all of history.* Perhaps the assembled crowd thinks the laws of nature will be suspended. Just as likely, they are out to gawk at the spectacle, suspecting that the government has foolishly underwritten a powerful man's idle dream.

From press accounts, official government documents, and the reminiscences of Langley and his colleagues, we know even the sounds and textures of the day's events: from the roar of the pathbreaking radial engine to the gossamer feel of the oiled white silk covering the aerodrome's skeleton. We know that the houseboat began to pitch as the wind picked up that afternoon. And we know that Langley, meticulous in his attention to detail, had prepared for almost every contingency. He had even contracted with Army surgeon Dr. Francis Nash to join the dignitaries on board the houseboat in case of a medical emergency.

Yet, for all the documentation and detail, two of the most important things about that December day remain elusive. A century later, we can only imagine the spectators' excitement: the suspense they felt as they stood for hours while Langley and his team readied the odd craft and their bewildered skepticism that, on this very day, a human being might actually unlock the secrets of flight.

Equally baffling, a fair accounting of the event's outcome is mired in controversy, ill will, and confusion. A century later, it remains one of the most compelling mysteries in the history of aviation. What, exactly, happened that day when one of the world's most eminent scientists tried to prove that a piloted airplane was possible?

The eighth of December arrived, unusually crisp, clear, and windless. Bad weather and dispiriting setbacks had repeatedly dashed Langley's hopes of conducting the final test of the new aerodrome and put his project far behind schedule.

Earlier in the fall, at a more remote spot forty miles south along the Potomac in Virginia, the team had attempted a series of mostly incomplete and unsuccessful preliminary trials. Langley favored the river for testing all his prototype aircraft designs. It offered an unobstructed expanse should the aircraft veer in flight; it was, in most spots, relatively shallow, further reducing the chance of losing the models; and, especially important with a human pilot, the water provided a more forgiving landing terrain than did solid ground. At least, that had been the thinking until the Potomac began to ice over.

Langley's houseboat laboratory was badly damaged in a mid-October storm, and, at the project's climax, weeks were lost while repairs were undertaken at a dock in Washington. Then, when the houseboat was finally restored, stormy, wintry weather prevented a definitive final trial of the machine. If these setbacks weren't enough, Langley later wrote, "more important and more vital was the exhaustion of the financial means for the work."

Taken together, technical difficulties, winter's onset, and funding worries have contributed to pervasive uncertainty and pressure for Langley and his coworkers. He knows that this clear December day offers one of the last, best chances to fly the aerodrome for months to come.

With the weather clear and calm throughout the morning, at noon Langley orders the crew to ready the aerodrome for an immediate trial flight. Charles Manly, his assistant and head engineer, will serve as pilot.

Langley's coworkers have been eager for just such an opportunity. But now there is much to ready and little time. The aerodrome sits in pieces inside the houseboat, and it needs to be carefully bolted together for the trial—a job that takes several hours. And the houseboat needs to be towed out from its dock at the end of Eighth Street in southwest Washington to a clear, wide spot on the Potomac where the flight can take place. Given the lack of notice, though, it is no small matter to commission tugboats for the operation. Finally, two tugboats—the *Bartholdi* and the *Joe Blackburn*—are hired for the job, but it is after two-thirty in the afternoon when they pull away from the dock with the large, flat-bottomed houseboat in tow.

For the sake of expediency, Langley decides to make the trial close to the city. This time of year, the Potomac is clear enough of traffic to permit it. Through blocks of floating ice, the two tugboats move downstream to the spot Langley specifies off Arsenal Point.

Because of the late hour, Langley also decides to forgo the task of mooring the houseboat. Instead, upon reaching their destination, the tugboats continue to run their engines against the current. Long cables from the tugs hold the houseboat moderately steady into a strong and gusty wind.

Throughout the morning, news of the impending flight has spread rapidly through rarefied Washington circles. Now, the spectators on the banks strain to see as the crew atop the houseboat roof struggles to assemble the aerodrome while the craft rolls and sways in the choppy water. The crowd can just make out the crew carefully placing the aircraft's two sets of wide white wings into position—the tandem-wing design that Langley came to favor from working for years with unmanned models. In between the fore and aft sets of wings are nestled the engine and seat for the pilot. Next, the workers fasten the aerodrome's Penaud tail: based on the design of the French aviation pioneer Alphonse Penaud, it looks from afar like an oversized version of the feathers on the end of an arrow or dart.

Those spectators with field glasses can clearly see Langley overseeing the affairs on deck, his earnest assistant and today's brave pilot Manly by his side. They can see Manly, a slight, energetic young engineer, quietly measuring himself against his mentor while excitedly helping to oversee affairs at the zenith of his youthful career. As these two men shuffle about, perhaps a few of the spectators can also make out their serious, even solemn countenance. But their nervousness and sense of urgency remain well hidden.

The two men do not discuss it, but the fact is, Langley and Manly have yet to succeed in their efforts to get the spring-loaded launching catapult atop the houseboat to work, despite numerous trials. Earlier in the fall, a full-blown failure had dumped Manly and the plane directly into the water. Yet they remain confident: they have repeatedly checked the launching mechanism and try to take comfort in the knowledge that Langley has used a similar method to great effect in his smaller-scale experiments.

In May 1896, for example, Langley successfully launched a thirteen-foot-long steam-powered model he called Aerodrome Model No. 5. This plane soared away from Langley's houseboat on the Potomac and motored along through the air for more than a half mile.

Far more than an early precursor to a working airplane, Langley's unmanned Model No. 5 was an unprecedented breakthrough that conclusively proved the viability of self-propelled, heavier-than-air flight. While balloons and dirigibles had successfully flown for some time, these were known, in the parlance of the day, as "lighter-than-air" craft because of their reliance on low-density gas or hot air to become airborne. Langley's successful flight of Model No. 5 was extraordinarily different and many immediately recognized its import.

Langley's astonished colleague and friend Alexander Graham Bell (by then already world-famous for his invention of the telephone) witnessed one of Langley's experiments on the Potomac River. As Bell soon told scientific colleagues, the model plane flew upwind above the Potomac until, exhausting its steam, it landed gently and unharmed on the water. The flight, he said, was "so steady that I think a glass of water on its surface would have remained unspilled." To Bell, the implications were enormous. No one present, he said, "could have failed to recognize that the practicability of mechanical flight had been demonstrated."

Through November of 1896, Langley repeated his success with an improved Model No. 6 that could fly up to three-quarters of a mile. And he took great pride in having crossed such a momentous threshold of aviation. As he proudly asserted in *McClure's*, a popular magazine of the day: "This has been done: a 'flying machine,' so long a type for ridicule, has really flown; it has demonstrated its practicability in the only satisfactory way—by actually flying, and by doing this again and again, under conditions which leave no doubt. "

More than seven years before the Wright brothers' success at Kitty Hawk, Langley's efforts spurred on the bicycle mechanics to conduct their own experiments. As Orville Wright put it later, Langley's flights were a key inspiration, emboldening him and Wilbur with the knowledge "that the head of the most prominent scientific institution of America believed in the possibility of human flight."

Ever the scientist, though, Langley was always more concerned with the laws of aerodynamics than the practical engineering challenges involved in building successful aircraft. Notably, then, he viewed the flights of Aerodrome Model No. 6 as something of an end in themselves. "I have brought to a close the portion of the work which seemed to be specially mine: the demonstration of the practicability of mechanical flight," Langley said, after the 1896 trials. "For the next stage, which is the commercial and practical development of the idea, it is probable that the world may look to others."

As it turned out, though, Langley would not leave the practical development of a full-scale flying machine to others because the War Department took careful note of his work. With the sabotage of the battleship USS *Maine* in Havana Harbor early in 1898, the United States had entered into the tempest of the Spanish-American War. Suddenly, planners at the War Department's Board of Ordnance and Fortification saw potential in the flying machine as a tool for military reconnaissance.

On the basis of Langley's stature and success, the military board, with President William McKinley's approval, recommended enlisting him on a secret project to develop a full-scale machine. Langley agreed to the plan. He insisted on independence and requested a budget of $50,000.

With the same methodical determination that had characterized his experiments throughout, Langley immediately set about to enlarge everything he had accomplished so far. Based on his success on the river with unmanned models, Langley decided to build a huge new houseboat that could serve as the base of a redesigned catapult launching pad.

Of course, the project was a hard secret to keep. Langley had widely publicized his earlier work with unmanned aeroplane models, and now he was building the Victorian-era equivalent of an aircraft carrier: 40 feet wide and 60 feet long, it would be large enough to contain the pieces of a full-scale airplane, a well-stocked workshop, and even sleeping quarters for a small staff. As one historical account put it, the hulking barge "easily took rank as the most remarkable thing in the way of marine architecture that had ever been seen on the Potomac."

If the enormous houseboat weren't odd-looking enough, the fifteen-ton iron superstructure on its roof drew the attention of even the most blasé. The massive launching structure, mounted on a large circular turntable, was designed to point the aerodrome directly into the wind after the houseboat had been anchored. Glimpsing the turntable and catapult, many who lived or worked along the river assumed that the massive and totally unfamiliar metal structure was the secret flying machine itself. And, as the *New York Times* later reported, the prevailing view—expressed repeatedly in bars and on roadsides—was unequivocal: Langley's machine would never fly.

Langley was for the most part unfazed by such opinions. Few people would say it to his face, but, despite the success of his models, the widespread view was that Professor Langley had gone off the deep end, squandering the Smithsonian's reputation and now the government's money chasing after a dream of cranks and charlatans. The disdain of his scientific colleagues was undoubtedly the most difficult to take; most held the view that playing with toy airplanes was not an endeavor worthy of a true man of science.

One highly respected colleague of Langley's, a Harvard-trained astronomer named Simon Newcomb, even published a high-profile scientific paper purporting to prove that powered "heavier-than-air" human flight was scientifically impossible. Newcomb's argument rested on the law of the cube: the sound principle that an object's weight increases at a dramatically faster rate than its surface area. As Newcomb noted, when a three-dimensional object grows in size, its volume increases by the cube, that is to say, by the product of all three of its spatial dimensions: height, width, and depth. But the surfaces that would actually support an aircraft in flight, being essentially two-dimensional, increase only by the square, that is, as the product of their two planar dimensions, height and width. The significance of this principle was that, while Langley might have been able to get a bird-sized model aloft, the task of building a full-scale airplane capable of carrying a person grew exponentially more challenging, if not entirely impossible.

Of course, Langley, who had experimented for years with a cleverly designed wind machine to shape his aerodrome's airfoils, knew that the key lay in the thrust of a powerful motor. But after scouring the United States and Europe, Langley was discouraged to find that no existing manufacturer could deliver an engine with the ratio of weight to power he needed. Internal combustion engines, a relatively new species at the turn of the century, were heavy: the best gasoline engines weighed upwards of twenty pounds for each unit of horsepower they delivered. To get the full-scale aerodrome off the ground, Langley knew, he needed at least a twelve-horsepower engine that was twice as light as the best engines available.

Unable to find such an engine, Langley decided to build his own. Polling his scientific colleagues to find the brightest new engineering mind in America, he engaged the services of Charles Manly, a promising engineer just graduating from Cornell University. By all accounts, Manly was a gifted, if unconventional, student. His habit was to abruptly leave his exams as soon as he thought he had done enough to earn a passing grade—much to the shock and annoyance of his teachers. Yet there was little doubt about Manly's command of the subject matter, especially of mathematical principles. When he entered Cornell as a sophomore in 1896, the math department professors determined that he was the best mathematics student ever to grace the university.

Upon joining Langley as chief assistant in 1898, Manly had no experience in the fledgling field of aeronautics, but he made good use of his exceptional math and engineering aptitude and delivered the final piece of Langley's grand plan. Although it would take years to complete, Manly created what was arguably the world's most advanced engine. Taking the most promising engine Langley had found,* a rotary engine that could produce only eight horsepower, Manly rebuilt it, crafting a five-cylinder, water-cooled motor that weighed 207 pounds (with 20 pounds of cooling water) but delivered more than 52 horsepower—an astounding ratio of less than 4 pounds per unit of horsepower. It was such a remarkable piece of engineering that no one would match it for almost a decade.

---

*An engine designed by a U.S. machinist named Stephen Balzer, who is best remembered for having built the first automobile to run in the streets of New York City.

Langley joined this engine to a full-sized aerodrome based upon his successful quarter-scale Model No. 6. Like the model, the plane used a tandem-wing design, with one set of wings in front of the engine and pilot and one behind. For stability, Langley placed the wings at a so-called dihedral angle: looking at the aircraft from the front, the wings protruded slightly upward from the fuselage, like a partly flattened letter V. By moving the craft's gangly Penaud tail up or down, the pilot would have vertical control, and the pilot could steer right or left by means of a small rudder placed below the second set of wings. The result was something like an overgrown mechanical dragonfly, a moniker the press had adopted for Langley's earlier unmanned models.

From tip to tip, the two sets of wings had a span of some 48 feet, giving the aerodrome more than 1,000 square feet of surface area to support itself—a total carefully calculated by Langley to take into account the law of the cube. In yet another notable innovation, his team constructed the aircraft's frame from steel tubing, making it both strong and light. Even including Manly and his engine, the entire craft weighed just 830 pounds.

On December 8, 1903, the years of effort seemed about to come to fruition. Langley had a tested design. He had the world's most advanced lightweight power plant, designed and built by his skillful young engineer. Now—with some luck—he would send a human being in a heavier-than-air contraption to soar like Icarus, albeit at a safer distance from the sun.

By 4:30 P.M., the winter sky is beginning to darken; shifting gusts make it exceedingly difficult for the tugboats, at the end of long cables, to keep the houseboat heading into the wind. In a hurried conference, Langley and Manly agree that they cannot postpone the test. It is a fateful choice. Perhaps Langley feels that his time is running out. As Manly recalls, the unspoken sense between them was that "it was now or never."

With excitement and resolve, Manly strips off his outer clothes. Since he knows that even a successful flight will land him in the frigid Potomac, he dispenses with the prevailing starched propriety of the day in the name of science: shedding his jacket and tie, he will make the journey in long johns, light shoes, and a specially made cork-lined jacket. In a charmingly optimistic gesture, Manly has also fastened a compass onto the left legging of his long johns. Presumably, it will help him keep his bearings in the event the machine carries him out of sight, far into the gusty sky before him.

Reaching between the plane's wings, Manly opens the throttle and cranks the propellers. The engine's roar silences the crowd onshore. Two mechanics remain on the upper deck, making final spot checks while Langley shepherds his distinguished guests onto small boats. According to Langley's plan, the boats will provide the dignitaries the best view of the launching—while allowing them to provide assistance should a rescue attempt be required.

Climbing through the aerodrome's brace wires, Manly seats himself in the cockpit. It is a small, three-foot-long, fabric-sided booth containing a wooden board to sit on and a single instrument: a tachometer to show the motor's speed. The tachometer's dial quickly registers that the motor has come up to full power, turning the two propellers behind his head at 950 revolutions per minute. He rests his hands upon two small wheels mounted before him. One controls the up-and-down motion of the tail; the other operates the rudder on the underside of the fuselage.

From his elevated perch atop the metal catapult, some sixty feet above the Potomac, Manly feels the houseboat rock and lurch beneath him. Ever a man of science, he focuses on the task at hand and tries to stay calm. The awed spectators watch as the white dragonfly stands still for a pregnant moment against the fading light. Finally, the young engineer raises his hand, signaling a mechanic to pull the trigger of the launching mechanism.

Perfectly, as planned, the aerodrome shoots along the rails of the launching track. It gains speed toward the track's end.

Suddenly, Manly hears a screeching sound and feels a violent jerk. He doesn't yet know it, but the tail of the plane bends sideways away from the fuselage. Thrust from the catapult's springs, however, nothing can stop the machine now, and it careens off the roof of the houseboat. With the plane's tail uselessly askew, Manly can't hold the aerodrome steady, and the powerful propellers push it nearly straight upward into the sky. For several long seconds, the aerodrome hovers like a helicopter before it slowly, inevitably, flips back on itself and crashes into the river.

Almost before anyone can register it, the machine is in the dark, cold water. It sinks below the surface, then bobs up to reveal a crumpled version of its former self. A sympathetic moan rises from the crowd as they watch for any sign of Manly.

Oddly, there will soon be a palpable sense of relief that the aerodrome doesn't fly—perhaps a kind of catharsis. While those on hand that day bemoaned the aerodrome's crash, public reaction to the news will be decidedly less supportive. In newspapers, in scientific circles, and in a rising tide of invective, the crash of the aerodrome is greeted not as a setback but as a deserved defeat—and a comeuppance for its backers. Almost uniformly, the press, academic colleagues, and government officials concur that Langley's failed attempt represents an appalling waste of money. A crazy idea. Even a national disgrace.

Press accounts of the fiasco greatly bolster public skepticism about the possibility of human flight. Ironically, in just nine brief winter days, on December 17, 1903, Orville and Wilbur Wright make history by carrying a man aloft for 852 feet along the windswept beach of Kitty Hawk, North Carolina. But, on this fateful cusp of aviation history, a *New York Times* editorial dubs Langley's project "Langley's Folly" and declares that the only way a "man-carrying" airplane might ever succeed is if mathematicians and mechanics work steadily at the task "for the next one to ten million years."

A few, like Alexander Graham Bell, defend Langley, but the swell of derision drowns them out. On the floor of the House of Representatives, for instance, Congress member Thomas Robinson of Indiana ridicules the expenditure of government funds on a flying machine. A regiment of Langley machines couldn't even "conquer the Fiji Islands," he argues, "except, perhaps, by scaring their people to death." Robinson likens the military's sponsorship of Langley's aerodrome to funding someone who promises to build "a railroad to the moon" or "buildings beginning with the roof, with no foundation." Many of Robinson's colleagues in the House concur. As Congress member Gilbert Hitchcock of Nebraska notes scornfully about Langley: "The only thing he ever made fly was government money."

Disappointed and demoralized by the failure and the storm of criticism, Langley turns his back on aeronautics. He resumes his duties at the Smithsonian Institution until his death, a little over two years later in February 1906. During those two years, he suffers often-merciless ridicule. Prior to the abortive flight, supporters had nicknamed Langley's plane the "dragonfly" for its tandem-wing design. Afterward, it is more often referred to as the "buzzard."

And what of Manly, the pilot and engineer? His own account captures the end of the ill-fated spectacle:

Almost before he could realize what was happening, Manly found himself submerged in the Potomac's icy December water, trapped beneath the aerodrome. As he recounts: "My cork-lined canvas jacket was caught in the fittings of the frameworks so that I could not dive deeper and get away. At the same time, the floor of the aviator's car was pressing against my head, preventing me from rising to the surface."

Onlookers watched nervously from the tugboats and other craft. But Manly remained trapped below.

"Exerting all the strength I could muster," Manly recounts, "I succeeded in ripping the jacket entirely in two, thus freeing myself from the fastenings which had held me. I dove under the machine and swam under the water for some distance until I thought I was out from beneath it."

Manly's ordeal was not yet over, though.

"Upon rising to the surface," he recalls, "I hit my head upon a block of ice. This necessitated another dive to get free of the ice."

One worker, Fred Hewitt, strained over the railing on the lower deck of the houseboat to try to glimpse Manly returning to the surface of the swirling, dark, deadly cold water.

"Upon coming to the surface," Manly recalled, "I noticed Mr. Hewitt, one of the workmen, about to plunge in. Before I could call out that I was safe, Mr. Hewitt had heroically dived in, thinking that I was trapped under the machine."

In the confusion, though, the houseboat had drifted and was now moving down upon Manly, Hewitt, and the aerodrome. Seeing this, Manly shouted for the tug to pull the houseboat away. And he and Hewitt were promptly yanked onto a rowboat by onlookers and ferried over to the houseboat.

Manly was uninjured but the frigid water had taken its toll; he collapsed as Dr. Nash hurriedly cut the clothing from his body. Moments later, wrapped in warm blankets and fortified with whiskey, the ever-courteous Manly startled the group by delivering what one discreet account describes as a "most voluble series of blasphemies." Dr. Nash said that, in his long career as a naval officer, serving in many parts of the world, he had never seen an act of courage "that equaled the cool valor of the pioneer pilot."

The onset of darkness made it difficult for workers to salvage the aerodrome from the river. By the time they finally hauled it back onto the houseboat, it was badly damaged and, having caught on the line fastened to the tugboat, had broken fully in half. Langley recalls that, in his years of testing aircraft, many accidents had caused equally serious damage to his prototypes. But, given the time of year, the lack of funds, and most of all the ignominious disgrace the team all felt, this accident would mark an end point for the aerodrome.

Nonetheless, Langley used his own personal funds to carefully box up the pieces of the aerodrome for storage at the Smithsonian. He said he was sure the parts might someday "attest to what they really represent as an engineering accomplishment."

Why did the aerodrome crash? Despite a century of speculation, the cause remains a mystery. We know the rear section of the machine dragged and collapsed before the aerodrome cleared the houseboat catapult. But what caused it to drag remains uncertain. Given a history of trouble with the catapult, extraordinary care had been taken to ensure that the mechanism would function flawlessly.

Major Montgomery M. Macomb, the official observer from the U.S. War Department Board of Ordnance and Fortification, leaves the question open in his official rendition of events. As Macomb reports: "The car was set in motion, the engine working perfectly, but there was something wrong with the launching. The rear guy post seemed to drag, bringing the rudder down on the launching ways, and a crashing rending sound, followed by the collapse of the rear wings, showed that the machine had been wrecked in the launching; just how it was impossible to see."

The wind could surely have been a factor. The day was clear, but the cold, gusty wind squalls added a considerable element of danger and uncertainty. Especially given the wings' broad surface area, an untimely gust could easily have blown the plane out of kilter as it sped down the launching track, causing it to snag on the mechanism along the way. In fact, it remains unclear why Langley and Manly decided to make the attempt at all that day given the unfavorable weather conditions. In retrospect, it seems like an uncharacteristic rashness overtook them after years of long and patient labor.

Of course, the leading explanation for the crash was that the aerodrome's frame was too weak or its design flawed. And this may have been the case. But Langley's record of accomplishment with similar designs and his systematic testing regimen—all recorded in copious scientific notes—cannot be ignored. The frame of steel tubing and the wooden-ribbed wings had all survived elaborate stress tests. Increasingly heavy bags of sand had been placed upon them to simulate air pressure, and great care was taken to make sure the wings' construction was both strong and flexible. As Manly put it:

"I cannot emphasize too strongly that there was neither fault in design nor inherent weakness in any part of this large aerodrome. The whole machine had been subjected to the most severe tests and strains in the [Smithsonian] Institution's shops in the endeavor to find any possible points of weakness and had shown itself able to withstand any strain it would meet in the air."

Today, of course, we would simply study the slow-motion replay or conduct a computer simulation to determine what went wrong. Certainly, even in 1903, a photographic record of the event might have helped solve the mystery of the flight's failure. But not a single photo captured the disabled craft at the onset of its cataclysmic failure. By 4:45 P.M. in December, the light had faded. The only known surviving picture of the flight itself, made by a *Washington Star* photographer, glimpses the aerodrome against a dark sky in a near-vertical position with its rear wings and tail in shambles.

Ultimately, all the second guessing cannot fully resolve the question of the aerodrome's viability—a question all the more dramatic and mysterious given its precedent-setting place in history. True to form, Langley limited his comments about the affair to the facts. He took pride, he said, that he and Manly had indisputably built a powerful engine twice as light as deemed possible by the best engine builders of the day. As for the aerodrome itself, he objected strenuously only to the "erroneous accounts" that asserted that the plane could not fly, noting that "the machine has never had a chance to fly at all, but that the failure occurred on its launching ways; and the question of its ability to fly is consequently an untried one."

While Langley was undoubtedly discouraged, turning away from the problem of flight at the end of his life, his younger colleagues and many admirers clung steadfast to the conviction that his "untried" plane would have flown if given the chance.

A decade later, when the age of aviation was in full bloom, a *New York Times* reporter looked back on his paper's earlier dismissive coverage and concluded: "The history of invention has no record more pathetic than that of Samuel P. Langley. At the very moment when success was in his grasp, when the dreams of a lifetime were about to come true and the labors of years of toil to be rewarded, the cup was dashed from his lips through the failure, not of the invention itself, but of a purely mechanical contrivance of minor importance. Derided in Congress and held up by the newspaper wits of the world as a target for their jests, Langley must have died a thoroughly discouraged man."

Alexander Graham Bell had arrived at a similar conclusion much earlier. After Langley's death, Bell delivered a eulogy at a commemorative ceremony: "No one has contributed more to the modern revival of interest in flying machines of the heavier-than-air type than our own Professor Langley." Bell added that his friend, as an aviation pioneer, had to face "not only the natural difficulties of his subject, but the ridicule of a skeptical world."

Ironically, years before the aerodrome fiasco, Langley had ended his well-known scientific textbook, *The New Astronomy*, with a parable that presaged much about the aeronautical predicament he faced at the end of his life. Langley intended the fable as an optimistic gloss on how little scientists know about the universe, but it stands equally well as an epitaph for a man who persevered for years amid a chorus of conservative critics and naysayers:

"I have read somewhere a story about a race of ephemeral insects who live but an hour," Langley wrote. "To those who are born in the early morning the sunrise is the time of youth. They die of old age while its beams are yet gathering force, and only their descendants live on to midday: while it is another race which sees the sun

decline, from that which saw it rise." In Langley's story, the generation of imagined insects alive at twilight gathers worriedly to hear
from their wisest philosopher about what to make of the imminent
sunset. The wise insect explains gloomily that generations of insect
scientists have determined through induction that the sun moves
only westward. And, since the sun is now nearing the western horizon, he says, "science herself points to the conclusion that it is
about to disappear forever," marking the certain demise of the
insects' world.

"What his hearers thought of this discourse I do not remember,"
Langley writes, "but I have heard that the sun rose again the next
morning."

PART I

REWRITING
AVIATION HISTORY

ONE

INTRIGUE AT HAMMONDSPORT

*If the Langley aerodrome flies, several chapters of
aviation history will have to be rewritten.*
—*BUFFALO EXPRESS* (BUFFALO, N.Y.),
MAY 20, 1914

The arrival of three imposing wooden crates has nearly halted work at the bustling Curtiss Aeroplane Company in Hammondsport, New York. It is a chilly afternoon early in April 1914 and, far upstate, spring has just begun to nudge the surrounding Finger Lakes region into bloom.

Workers haul the huge, pine-planked boxes, one by one, into the open courtyard outside the company's collection of gray hangars. As they do, more than half of the plant's one hundred employees stream outside to get a better look. Crates of parts, tools, and equipment arrive at this airplane factory almost every day. But today's boxes—sent by rail from Washington, D.C.—are an unprecedented delivery, the subject of hushed gossip at the plant for weeks.

Henry Kleckler, the shop foreman, wipes the grease from his hands and steps into the courtyard to help as his boss, Glenn Hammond Curtiss, approaches the largest crate. Curtiss is tall and trim, with a reserved intensity. He is just thirty-six years old, but his thinning hair and serious countenance give him an ageless air of authority. He is also a corporate executive more comfortable on the shop floor than in a boardroom. His easy rapport with his workers is obvious in the way they enthusiastically surround him.

Now, as Curtiss pries off the crate's big wooden top with the back of a hammer, the crowd of assembled mechanics, carpenters, and engineers falls silent. Inside the box lie the crumpled wings of the most maligned airplane of all time: Samuel Langley's aerodrome, his infamous seminal attempt to create a piloted, heavier-than-air flying machine.

The first peek is not encouraging. Packed over a decade ago according to Langley's instructions, the contents appear a terrible mess, full of twisted metal, broken wood, and tattered fabric. But as the knowledgeable workers draw closer to inspect the pieces, their initial dismay turns to admiration. Though old and badly damaged, the antique machine's craftsmanship is unmistakable. The wooden ribs of the aircraft's wings are not only exquisitely joined; they have been hollowed out to make the craft lighter. Unlike the canvas muslin used on most modern airplanes in 1914, the wings of Langley's plane are sheathed in a fine skin of now-rotted, oiled silk. Curtiss calls it the most beautiful piece of work he has ever seen.

Now the hard part must begin. At the behest of the Smithsonian Institution, a team at the Curtiss plant will try to restore the machine to its original condition. The goal: to see whether, if properly launched, Langley's plane can fly.

Confronted by the remains of the aerodrome, the workers recognize the scale of the painstaking restoration before them and wonder skeptically whether the battered and unconventional-looking machine will ever get aloft. Focused on the immediate problems of reconstruction, they are all but blind to the broader implications of tampering with the judgment of history. No one present realizes that before they are through, their efforts will ignite one of the most bitter controversies in the annals of aviation.

How strange are the whims of history and how difficult to predict and understand. Few could have expected the extent of ridicule Langley suffered for the aerodrome's failure or that, after languishing for more than ten years in the back of a carpentry shop at the Smithsonian Institution, the crumpled aircraft would once again become the subject of intense interest. Fewer still could ever have foreseen the aerodrome's voyage to this unlikely destination, rural and remote, some fifty miles southeast of Rochester, New York.

For generations, the Finger Lakes region has been known as New York State's wine country, home to hundreds of acres of vineyards nestled among tree-covered hills that slope to the edges of a series of long and narrow freshwater lakes. In this bucolic area, Mark Twain spent most of his summers and wrote some of his best-known works, including *Tom Sawyer* and *The Adventures of Huckleberry Finn*. The towns here exude an upright American charm; peaceful, but too industrious-seeming to feel sleepy. In the heart of the region, the small town of Hammondsport is no exception, with a postcard village square lined with substantial brick storefronts.

Strangely enough, since the earliest years of the twentieth century, this improbable spot, far from any major metropolitan area,
has seen a bustle of activity that will forever mark it as the "cradle of
aviation." In fact, by 1914, Glenn Curtiss has amassed in Hammondsport the best and largest collection of skilled aircraft
mechanics to be found anywhere in the world. As a result, the
town's residents have never felt so much at the center of things as
they do now. Over the past several years, it seems that everyone
with an interest in airplanes—from inventor Alexander Graham Bell
to industrialist Henry Ford—has made their way here to the Curtiss
Aeroplane Company.

"Everybody in Hammondsport has an expert's familiarity with
aeroplanes," gushes a reporter from Joseph Pulitzer's *New York Sun*
on assignment to Hammondsport in the spring of 1914. "The most
astonishing experience of the visitor is to hear an eight-year-old
child talk about the virtues of flat surfaces as compared to curved
surfaces with the glib sureness of an expert," he writes, "or to
engage a charming young woman in conversation . . . and have her
give a learned dissertation on the thrust of propellers."

The catalyst for all this interest, the magnet for all this excitement and industry, is the quietly irrepressible Glenn Curtiss.
Despite his relative anonymity today, Curtiss surely belongs in the
pantheon of America's greatest entrepreneurial inventors. With
uncanny regularity, his remarkable career led him to the heart of
some of the most important pioneering developments in the history of aviation. In the course of a few short decades, Curtiss
arguably contributed more to the modern airplane than anyone
before or since, including: the first public flight in the United
States, the first commercially sold airplane, the remarkable first
flight from one American city to another, the issuance of the first

U.S. pilot license, to name just a few momentous breakthroughs. Ask almost anyone today and they will likely tell you that these milestones were achieved by the Wright brothers, the legendary team that most of us—nearly a century hence—immediately associate with the dawn of aviation. However, the first public flight in America was not made by the Wrights, whose obsession with keeping their research secret shrouded the early years of aviation in internecine intrigue and legal wrangling. Rightful claim to all the above achievements belongs to Curtiss.

Like the Wright brothers, Curtiss ended his formal education in the eighth grade. But despite the lack of schooling, historians credit Curtiss with a central role in no fewer than five hundred aviation innovations. Even more impressive, many of his seminal contributions are still in use in airplanes today, including everything from wing flaps and retractable landing gear to the enclosed cockpit and the design of the pontoons used on seaplanes. By contrast, virtually none of the Wright brothers' aeronautical designs has stood the test of time. Most of the Wrights' practical engineering contributions were obsolete by as early as 1912.

If Curtiss's phenomenal creativity was exceeded by anything, it was by his energy and drive. His competitive spirit was evident from the start of his career. A national bicycle champion by the age of twenty, he went on to win world renown as "the fastest man alive" by riding a motorcycle of his own design at a record-breaking 136 miles per hour in 1907. Curtiss's talent and daring proved a formidable combination and he would need both to make his greatest contribution: opening the sky to the commercially viable modern airplane.

First and foremost here in Hammondsport in 1914, though, Glenn Curtiss is a local hero. Indeed within a radius of a hundred miles around Hammondsport, Curtiss and his aeroplanes have become an almost ubiquitous topic of conversation. Everybody seems to feel a proprietary interest in his career. His portrait even hangs in the local post office, as it does in the window of Hoyt's pharmacy, with the caption "He's good enough for us."

To some extent, Curtiss has been a local favorite since he was a boy. He was an enthusiastic and energetic child and, sadly, his father and his grandfather, both Hammondsport residents, had died by the time Curtiss was four years old. As a result, many in town took a special interest in him. Jim Smellie, a local shopkeeper and friend, helped coach Curtiss as a bicycle racer. And at a key juncture in Curtiss's career when he was twenty-one, Smellie, who was expanding his general store, offered to turn over to Curtiss his bicycle repair and spare parts business. On the spot, Curtiss decided to open a bicycle shop in Hammondsport. Mrs. Malinda Bennitt, a wealthy widow in town who had always been fond of him, gave Curtiss rent-free use of a narrow storefront on the town square to help him get a start in business.

Hammondsport would never be the same.

Building upon his passion for bicycle racing, Curtiss moved quickly from repairing bicycles to building and selling them—and then to experimenting with the novel idea of motorized cycles. Locals could hear him for miles around careening wildly along dirt roads on the outskirts of town on an early belt-driven prototype he called the "Happy Hooligan."

Curtiss learned remarkably quickly from his early experiments. Soon he was not only building motorcycles and lightweight motor-cycle engines but filling orders for them from across the country. From the moment he started his own manufacturing outfit, tales of Curtiss's inventive spirit became part of the lore of the town. One day, as one of these many stories goes, Curtiss idly twisted a rubber grip on the handlebar of a motorcycle while he was standing outside talking with a customer. Abruptly halting the conversation, Curtiss ran into the shop, beckoned a mechanic, and on the spur of the moment invented the handlebar throttle control—a design that would, of course, become a signature feature on almost all modern motorcycles.

Within a few years, Curtiss transformed his modest bicycle shop into an impressive manufacturing operation, turning out a wide array of motorcycles, lightweight engines, and ultimately aircraft. By 1914, he reached a wholly new phase of prominence. The Curtiss Aeroplane Company not only had two large plants and more than one hundred workers; it occupied a place of international stature in the emerging aviation industry.

Nonetheless, as Curtiss biographer C. R. Roseberry notes, Curtiss always referred to his operation as "the shop." The word "factory" was simply not his vocabulary. For Curtiss, his shop was a place in which workers collaborated, openly exchanging ideas for modifications and improvements. It was also a place where Curtiss could always dirty his hands alongside his employees to try out new ideas whenever he felt so inclined.

To help to ensure his vision of a vibrant, creative workplace, Curtiss surrounded himself with both close childhood friends and exceptionally skilled mechanics. Plant manager Harry Genung had been one of Curtiss's best friends since grade school. Genung was affable, organized, and unfailingly loyal to Curtiss in a relationship that extended far beyond the confines of the rapidly expanding business. For years, Genung and his wife, Martha—who also worked at the "shop"—even lived in the back of the same big house where Curtiss resided with his wife Lena.

Meanwhile, Henry Kleckler, Curtiss's shop foreman, earned a

near-legendary status among early airplane designers for his natural engineering gifts. As one account puts it, his coworkers used to say that Kleckler, a thickset man of Dutch descent with little formal education, could make a motor out of a piece of baling wire. They would call for him whenever they were stumped by a difficult piece of work. Kleckler would arrive grinning, listen intently to what was required, and invariably say, "I fix him." No matter how big his company grew, Curtiss maintained: "I'd rather have Henry Kleckler on a project than six engineers."

Like many of the employees, Kleckler and Genung had joined up with Curtiss early in the company's phenomenal period of growth. In 1909, when A. P. Warner—the first private individual to buy an airplane in America—came to the plant to make the sale, he remembers Curtiss's operation as "little more than a shed with a few tools in it." As the company rapidly expanded over the next five years, Genung and Kleckler were indispensable to Curtiss. When Curtiss got swept up in a new idea, he needed Genung to manage an often overwhelming volume of existing business. And to bring his new ideas to fruition, Curtiss invariably relied on Kleckler to work out the knottiest engineering details.

But the two men were surprisingly representative of the devoted and genial workforce that Curtiss had assembled by 1914. Of all the many employees who worked for him over the years, few felt differently than Lewis Longwell, who joined the company in 1911. As Longwell puts it simply, "Curtiss was a good and honest man to work for." A key secret to the company's success, another Curtiss collaborator, Theodore "Spuds" Ellyson explains, was that Curtiss never set himself apart as a genius inventor. Rather he was "a comrade and chum, who made us feel that we were all working together, and that our ideas and advice were really of some value."

• • •

Examining each piece of the Langley aerodrome carefully as it is uncrated, Curtiss and Kleckler personally supervise as workers carry the pieces into a newly created work area off the courtyard at the Curtiss plant. The outdoor space between the plant's office building and the airplane assembly room has recently been walled in and roofed over and it will serve as the staging area for the months of work it will take to restore the aircraft.

It is an indication of the stature of the aerodrome restoration project that Curtiss has put Kleckler in charge. But, uncharacteristically, Kleckler is worried and, as the two quietly discuss the job, he doesn't hesitate to say so. Chief among Kleckler's concerns is that no expense was spared in the aerodrome's original construction and he doesn't know how he and his workers will be able to match it without spending a fortune far beyond the limited budget the Smithsonian has proposed for the job.

Curtiss, as always, is optimistic. They can make the wings' ribs out of solid spruce, he assures Kleckler. The spruce will be heavier than the fine, hollowed-out hardwood, but the pieces will be much cheaper and easier to construct. And the same kind of corners can be cut on the plane's skin, he says. Silk would be lighter, but it is just too expensive; they will have to make do with the canvas the shop already has on hand.

As Curtiss explains, Professor Albert Zahm from the Smithsonian will be arriving soon to oversee the work and make sure it follows Langley's original specifications. They can let him and Smithsonian Secretary Charles Walcott weigh in on those kinds of issues. But, he adds, it is unlikely that they or anyone will be concerned by simple substitutions for cheaper and heavier materials. The key is simply to alter the plane's design and aerodynamic properties as little as possible.

Inside the new staging area, the workers hang the main piece of the aerodrome's fuselage from the roof rafters to inspect it more closely. Despite the damage, the aerodrome's steel frame is still clearly recognizable. The workers can still visually trace the aerodrome's general system of control: from the pilot's seat under the plane's backbone, the lever is still there to move the now-crumpled tail up and down, as is the one designed for steering right and left, although only a broken fragment of the vertical rudder now remains.

Almost certainly, before he leaves the shop, Curtiss announces to the group his conviction that the aerodrome will fly. But, like everyone present, he knows that the proposition is highly uncertain, more of a wish than a real assessment. Hanging from its new perch, the hollow steel carcass of the aerodrome looks like it might be more at home in the Smithsonian's collection of fossilized dinosaur skeletons.

In the first weeks of work the biggest revelation for Kleckler and his team is the engine. Although it has been thoroughly waterlogged and subjected to more than a decade of neglect, there is no question about the sophistication of its design. Workers repeatedly marvel at its resemblance to the new aircraft engines the Curtiss plant builds. By 1914, the Curtiss Aeroplane Company, like most modern aircraft manufacturers, has shown a preference for radial designs, similar to the one pioneered by Manly, in which the engine's five cylinders are arranged around a central hub like a bulbous five-pointed star. Manly's means of cooling the engine was also ahead of its time, employing a water cooling system to help dissipate the intense heat generated by the engine.

Before long, Charles Manly himself, Langley's close assistant and chief mastermind behind the aerodrome engine, will make the pilgrimage to Hammondsport to lend a hand rebuilding the aerodrome. Manly has long since gone on to design and manufacture trucks with hydraulic controls, but he is understandably elated that the aerodrome will finally get another chance to prove itself airworthy. Like so many others, Manly has spoken often of retesting Langley's machine. He brought up the issue in 1908 with the Smithsonian when he became one of the founders of the Aero Club of America. And he noted publicly in 1911, when Langley's papers were posthumously published, that he hoped eventually to raise the funds to do the job himself.

Even with Manly's help, though, Walter Johnson—who worked closely with Kleckler on the project—recalls, they could never keep the motor hitting on all five cylinders. The team gave the engine a hotter spark, using the more up-to-date magneto ignition instead of the previous dry-cell batteries, but for all their efforts, they could only get the motor to develop two-thirds to three-quarters of the horsepower it had originally demonstrated. A decade of rust after its immersion in the Potomac had simply taken too big a toll for it to be brought back to its original condition.

The team members know that the lack of horsepower will greatly jeopardize the plane's chances of getting aloft. But they are determined to fly the aerodrome with Langley's original engine. Ultimately, only the engine's carburetor will be changed, and that only because Kleckler can't understand the workings of the original mechanism well enough to try to restore it. But even the worries about the sufficiency of the motor's now-diminished power don't change the high esteem Kleckler and his team hold for Manly's superb piece of engineering.

The farsightedness of Manly's engine design is all the more remarkable because so much else had changed since he first built it.

While the aerodrome was collecting dust in storage at the Smithsonian Institution, a full-blown technological renaissance had transformed the world. In fact, in the course of human history it would be hard to find a more eventful decade of dramatic technological change than the one Langley's aerodrome had quietly missed while boxed up in that back room.

The first decade of the twentieth century saw the advent or ascension of a host of revolutionary technologies, including the telephone, electric lighting, and power transmission, the automobile, the radio, the phonograph, and motion pictures. People were staggered by the changes and the world had never seemed so full of possibilities or so unsettled by them. Inventors like Edison, Dow, Deere, Westinghouse, and many others laid the foundation for a new corporate America, which, in that fateful first decade of the twentieth century, quietly began to surpass all other nations in the production of most tangible commodities from coal and chemicals to steel and heavy machinery.

Technology's latest offerings were cropping up everywhere. Henry Ford's Model T, introduced in 1908, had begun literally to make over the landscape at home while, by 1914, the opening of the Panama Canal reshaped it on an unimaginably vast continental scale. It seemed as though there was almost nothing that technology couldn't accomplish. And the field of aviation was as good an example as you could find.

Perhaps nothing illustrated the pace of change better than this single fact: One September day in 1913, just months before the aerodrome's arrival, a pilot named Link Beachey strode into the Hammondsport plant and asked Curtiss to build him an airplane that could "loop the loop." Beachey, indisputably the greatest and most fearless stunt aviator of the day, wanted an engine "more than twice as strong as any of the standard makes," whose gas flow wouldn't be cut off by flying upside down. As part of the package, he specified that the plane be built with a special harness that would fasten him in for the stunt. The ever-prudent Curtiss was reluctant, viewing the idea as brazen and needlessly risky. But Beachey, the equivalent of a modern-day rock star or Hollywood celebrity who drew huge and enthusiastic crowds to all his venues, was a hard man to decline.

By Thanksgiving, in a special plane Curtiss built for him, Beachey was astounding huge crowds as the world's first pilot to fly multiple loops in the air.*

All of which might well lead one to wonder, if Curtiss could so handily build a plane that could loop the loop, why would he ever be so keen to try to rebuild Langley's ancient and outmoded aerodrome?

The answer, at least on one level, is that, by 1914, the youthful aviation industry was undergoing an identity crisis. The strange, unresolved saga of the aerodrome had left a persistent and nagging question about how the history of the airplane should be told. And with some of the field's doyens reaching the end of their lives, the question was one of more than idle import. Almost from the time of the aerodrome's crash, for instance, Alexander Graham Bell had called for resurrecting the aerodrome. Octave Chanute, another of Langley's contemporaries and one of the earliest aviation researchers in the country declared in 1909, for example: "There is no doubt that if the [Langley] machine had been properly launched it would have flown. The machine is still in existence," he noted,

---

*Much to Beachey's chagrin, however, the French aviator Adolphe Pegoud would go down in history as the world's first aviator to successfully accomplish a single airborne loop on September 21, 1913.

calling it "most unfortunate that further effort had never been made" to test it.

Meanwhile, the seed for Curtiss's involvement in the project was planted in 1913 when the Smithsonian Institution awarded him the Langley Medal, the nation's highest aviation award. The award had been inaugurated in 1908, several years after Langley's death, and presented to the Wright brothers for their successful pioneering flights at Kitty Hawk. Since then, no achievement in aviation had been deemed worthy of the honor until the board voted to recognize Curtiss for his recent invention of the hydro-aeroplane, now known as the seaplane.

On May 6, 1913, at the ceremony for the presentation of the Langley Medal, Alexander Graham Bell made the lengthy tribute to Curtiss before a Washington, D.C., audience, which included a diehard contingent of aviation buffs dedicated to Langley's memory. Among the old gang of Langley supporters were Charles Walcott, who succeeded Langley as secretary of the Smithsonian; General James Allen, president of the Aero Club of Washington; Samuel B. McCormick, chancellor of the University of Pittsburgh, where Langley had held a chair in astronomy before coming to Washington; and Langley's close former assistant Dr. John A. Brashear, who was given the honor of unveiling a tablet dedicated to Langley at the event.

"I simply wish to express my feelings of gratitude and pleasure," the always shy Curtiss remarked upon accepting the medal—a medallion fashioned from a pound of solid gold. Even at this stage in his career, Curtiss was uncomfortable making speeches. But he had quickly picked up on the tenor of the evening, adding, "As I look at the Langley models here, it becomes more evident to me than ever before—the merit of these machines and the great work which Mr. Langley did." Ending his remarks to a flood of applause Curtiss noted, "I cannot say too much in favor and in memory of Professor Langley."

There is no evidence that the idea to rebuild Langley's plane was formally hatched at this event, but there is little doubt that the notion crossed many minds before the evening was over. Like the rest of the audience, Curtiss was doubtless moved by Brashear's almost maudlin lament about the last half hour he spent in Langley's office. Brashear recalled that Langley had shown him a small piece that had broken off from the aerodrome's launching mechanism that he believed had foiled the fateful attempt. "With a sad heart he turned to me and with trembling voice said, 'Mr. Brashear, this has wrecked my hopes forever. My life work is a failure.' I did all in my power to cheer and comfort him, but it was too late.

"Soon after that," Brashear continued, "he passed away, and I have often—aye many, many times—thought of that last sad half-hour spent with him. He was a noble man, and his works, though suddenly cut off by death, will live as long as this old world shall have dwellers upon it."

If remarks like Brashear's brought Langley's sad tale back to the attention of many in the field of aviation, the opportunity to restore the aerodrome was greatly enhanced by the Smithsonian's recent establishment of a new department called the Langley Aerodynamical Research Laboratory. Headed by Dr. Albert F. Zahm, a noted aeronautical scholar of the period, the new lab was designed to spearhead American theoretical flight research.

Whereas Langley's many supporters viewed a restoration of the aerodrome as a means to reclaim his reputation for posterity, Zahm hoped the aircraft might also help to establish the standing of his fledgling laboratory. In particular, the past several years had seen an alarming number of fatalities. At least eight aviators had been killed in highly publicized crashes in 1913 alone. Zahm had noted that most of the crashes occurred when the planes lost fore and aft stability, leading to a sudden, uncontrolled dive. He believed that his new lab could help the aviation industry by reviewing the tandem principle adopted by Langley to see whether Langley's design might "lessen, if not entirely prevent a fatal dive."

In interviews in Hammondsport after Zahm's arrival, Curtiss explains his undoubtedly heartfelt admiration for Langley. And he echoes Zahm's hope that experimenting with Langley's tandem-wing design might, as he put it, "affect the form and structure of aeroplanes" in the future. But Curtiss has another reason to try to resurrect the old aerodrome—a reason so urgent and explosive that it will, as his friend and colleague Lyman Seely puts it, ultimately help spawn "the most persistent and the most misleading propaganda ever attending a scientific test."

WRIGHTS AND WRONGS

*In Hammondsport, the old-timers used to say that if
you jumped up in the air and flapped your arms you'd
be infringing on the Wrights' patent.*

> —TONY DOHERTY, SON OF
> CURTISS'S ASSISTANT
> ELLWOOD "GINK" DOHERTY

By the time the aerodrome arrives
in Hammondsport in the spring of 1914, Glenn Curtiss faces an
extraordinary situation. He has won almost unanimous admiration
from practitioners in aviation around the world. His airplanes have
broken distance, speed, and altitude records. But in January, the
U.S. Court of Appeals has just handed down a permanent injunction that prohibits Curtiss from manufacturing or even exhibiting
his aircraft in the United States without a license from Orville
Wright. And, despite Curtiss's repeated attempts to negotiate,
Wright has announced that he will consider lenient royalty arrangements with anyone in the field *except* Curtiss.

In a startlingly broad interpretation of the Wright brothers' patent, the courts have sanctioned their exclusive claim to the sole practical means of stabilizing an aircraft in flight. As the Wrights had hoped, their sweeping patent has become, in effect, a patent on the airplane. And especially since the death of his brother Wilbur in 1912, Orville Wright is in no mood to compromise: he unwaveringly demands 20 percent of the revenue generated by any competitors' airplanes whether through their sale or exhibition.

As a result, unless Curtiss decides to move his company to another country that does not recognize the Wright patent claims, he will be forced to either cease his operations or pay such a crippling mountain of back royalties on the planes he has already sold or flown that he will surely be bankrupted.

There is also no question that the feud has become personal. In a front-page interview in the *New York Times* in February 1914, Orville accuses Curtiss of stealing the Wrights' designs and even blames him for Wilbur Wright's death from typhoid fever. According to Orville, Wilbur's agitation over the case "worried him to his death . . . first into a state of chronic nervousness, and then into a physical fatigue which made him an easy prey for the attack of typhoid which caused his death."

Calling Orville's claims "absurd, if not malicious," Curtiss publicly retorts that he never "had an item of information" from the Wright brothers that helped him build his airplanes. As for the contention about Wilbur's death, Curtiss describes it as a bunch of "insinuations easily interpreted as such untruths as I cannot believe Mr. Wright, or any other sane man, ever made."

Curtiss has repeatedly appealed to the Wrights for a settlement. With no success in negotiations and no satisfaction in court, he has few remaining options. In a move born of desperation, Curtiss embarks on the aerodrome restoration as an ambitious effort to set the record straight.

The Smithsonian's involvement in restoring Langley's aerodrome lends an aura of objectivity but, in 1914, the question of Langley's contribution to aviation is of more than academic interest to Curtiss. He knows that if Langley's plane flies, it could raise profound doubts about the Wrights' claim to being the first with a "useful" aeronautical invention.

In fact, Curtiss is quite candid about his motives. As he writes to Link Beachey, another target of the Wrights' patent claims, the aerodrome restoration "would go a long way toward showing that the Wrights did not invent the flying machine as a whole but only a balancing device." If they are successful in getting the aerodrome to fly, he says, perhaps in court "we would get a better decision next time."

Today, nearly a century into the age of aviation, the Wright brothers have become a part of our collective mythology—lone inventors who single-handedly turned a fantastic dream into a practical reality. Of course, the myth captures a truly monumental achievement. But it also willfully ignores half of the Wright brothers' story, obscuring the role the Wrights played once their invention took flight.

Now largely forgotten, the Wrights made no secret of the fact that they sought a monopoly on production of the airplane comparable to the one Alexander Graham Bell had won for the telephone. After all, monopoly was the hallmark of the Wright brothers' era—the Gilded Age—with vast, vertically integrated empires of oil and steel built by titans like Rockefeller and Carnegie. Securing monopoly control of the commercial airplane was the linchpin of the Wrights' business strategy. It helped them to attract a bevy of heavyweight backers that came to be known in many circles as the Wall Street Air Trust, including powerful financiers like Robert Collier, August Belmont, and Cornelius Vanderbilt.

With the help of Vanderbilt and the others, the Wright brothers came close to achieving monopoly control over airplane production in the United States through broad patents and aggressive business tactics. But unlike Bell's phone lines, which were conducive to a centralized monopoly, the chaotic, creative drive to conquer the sky in the first decade of the twentieth century would prove exceedingly difficult for the Wrights to contain. And the biggest obstacle they faced was their most formidable competitor: Glenn Curtiss.

After the appeals court decision in 1914, Orville Wright boasted to the press that the Wright Company had secured "absolute control" of the emerging airplane industry. His remark was delivered with a shrewd eye toward his company's investors but a remarkable blindness to its effect on the industry.

The truth is, by this time, the Wrights' handling of their proprietary rights over the course of nearly a decade had already alienated most of their colleagues in the young aviation field. Even Grover Loening, a loyal friend to the Wrights and onetime chief engineer of the Wright Company noted later that, by filing suit—and, in particular, by prosecuting their case so aggressively against Curtiss—the brothers "turned the hand of almost every man in aviation against them."

The Wrights' legal case against Curtiss hinged on a particular technical issue. The Wright brothers had solved the difficult problem of stabilizing an aircraft by making the wings of their planes flexible. In their patented "wing warping" method, the Wrights twisted the airplane's wings in a system wired to the plane's rudder. When the Wrights steered their airplane, their wing-warping system twisted each wing slightly in the opposite direction to help the plane bank on turns without losing control. Although some of the most respected aviation pioneers claimed the underlying principle was well known for as many as fifty years before the Wrights, their wing-warping stabilizing device was nevertheless an important advance that helped the airplane achieve controlled flight.

From the first plane Curtiss ever built, he and his team solved the stability problem in a related, but notably different way. Realizing that the function of the rudder and the need for lateral stability were separate, they designed wing flaps, or so-called ailerons.* Like the Wrights' technology, the ailerons on each wing tilt in opposite directions to stabilize the aircraft. Unlike the Wrights' design, however, the ailerons operate separately from the plane's fixed wings and from its rudder. Ailerons rapidly became the industry standard. With them, an airplane's wings could be made rigid and much stronger, and they allowed the plane to remain stable independent of its steering mechanism.

Curtiss and most other aviators of the day argued that the aileron was a significant and distinct advance that should not be legally covered by the Wrights' claims. In essence, though, the Wrights claimed that their completion of the first proven invention to solve the problem of lateral stability gave them rights to any subsequent design, including ailerons. The Wrights' patent itself repeatedly spells out this contention: "We do not wish to be understood as limiting ourselves strictly to the precise details of construction herein-

---

*The term *aileron*, French for "little wing," is attributed to aviation pioneer Robert Esnault-Pelterie, who experimented with them on a full-scale glider in 1904.

before described," the Wrights' lawyers write at virtually every point in which the patent spells out the particulars of the brothers' wing-warping technique. "We do not limit ourselves to the particular description of rudder set forth . . ." it reads when they describe the rudder. And, importantly, when they describe the way the wings can be flexed, the patent states: "Our invention is not limited to this particular construction."

With such all-encompassing language passing muster at the U.S. Patent Office, the Wrights' lawyers argued that it mattered not that wing warping was virtually obsolete within six years from the time the patent was issued. Nor was it relevant that even the Wright Company would quietly begin to abandon wing warping in favor of ailerons by as early as 1915. As their lawyers argued, the Wrights had been granted exclusive rights to all known means to laterally stabilize an airplane; now Orville Wright could legally exercise this exclusive proprietary claim however he wished.

It is a detail lost to history exactly when Orville Wright first learned that Curtiss and the Smithsonian team intended to restore Langley's aerodrome. Most likely, he read it in his morning newspaper as did other Americans early in the spring of 1914.

We can imagine him at home with his older sister Katharine and his father, Bishop Milton Wright, in their new mansion on Hawthorne Hill, southeast of Dayton, Ohio, reading the news, his cheeks flushing with rage. Not one to easily voice his anger, Orville might well have risen from the breakfast table in a wordless fury to pace and fume outside on the grand porch above his stately, sloping lawn.

Or perhaps not.

The particulars of the scene may have been forgotten, but we do know something about the degree of Orville's distress over the aerodrome affair. Legal ramifications aside, he took the restoration of Langley's airplane as a personal affront and believed that Smithsonian Secretary Charles Walcott was colluding with Curtiss in a plot to steal his and Wilbur's rightful claim to being the first in flight. In the coming weeks and months, a wide variety of visitors to Hawthorne Hill note Orville and Katharine's agitation over the matter with some alarm.

According to historian C. R. Roseberry, Ohio governor and Wright family friend James Cox was taken aback with the vociferous way the usually demure Katharine denounced Curtiss's aerodrome project as "a fake . . . so raw that it seems incredible." Holden C. Richardson, who would go on to become a captain in the U.S. Navy, was an overnight guest in the Wright home during this period. "Katharine especially was terribly bitter toward Curtiss," Richardson remembers, and couldn't seem to forgive him. Moreover, he recalls, because he was known to be a friend of Curtiss's at the time, Katharine had difficulty treating Richardson himself with civility.

Grover Loening, then chief engineer of the Wright Company, worried that Orville and Katharine's dislike of Curtiss was getting the better of them, "preying on their minds and characters." The aerodrome restoration, he says, became a "great hate and obsession" in the Wright household, "It was," Loening recalls, "a constant subject of conversation, and the effort of Curtiss and his group to take credit away from the Wrights was a bitter thing to stand for."

No one would take the matter more to heart, however, than Griffith Brewer.

Brewer was an English attorney who had met the Wrights in Europe, had helped them secure financing for the British Wright

Company, Ltd., and had fought to protect their patent rights and collect unpaid royalties in England. During the spring of 1914, Brewer was invited to spend three months as a guest at Hawthorne Hill. He had recently signed a contract to write a book on the emerging aviation industry. But, given the timing of his stay, it is not surprising that his project was waylaid. Instead, Orville convinced Brewer to go check up on the Langley restoration project. As Brewer would note later, Orville dispatched him on a mission "to go to Hammondsport and find out what Glenn Curtiss was doing to falsify the history of aviation."

Brewer made the trip, writing later that he felt "like a detective going into hostile country, where I should get rough handling if my mission were known." He never let on who he was, but he managed to catalog an impressive laundry list of changes Curtiss's team was making to the aerodrome. He then wrote a prominent letter to the *New York Times* about the case. Many of the entries on Brewer's list of technical objections seem petty, such as his contention that the Curtiss team had installed a modern carburetor or that the aircraft's Penaud tail had been positioned twenty inches higher in the rebuilt version than in the original. But Brewer did successfully cast doubt on the motives behind the experiment. In the most obvious and stinging of his accusations, Brewer asked: "Why, if such a demonstration were decided on, was not some impartial, unprejudiced person chosen to make the tests, instead of the person who had been found guilty of infringement of the Wright patent?"

Brewer's complaint resonated because, indeed, the team restoring Langley's aerodrome did have serious conflicts of interest. But rightly or wrongly, Zahm and Walcott always considered themselves to be impartial observers of the operation under the auspices of the Smithsonian. And, as Brewer undoubtedly knew, given the state of the industry at the time, there were few other candidates as qualified as Curtiss to undertake such a restoration. Zahm's fledgling aerodynamical laboratory was not equipped with the personnel to do the job and the Smithsonian would certainly have had no luck appealing to the Wright Company. Furthermore, aware of their delicate position, Curtiss and his team allowed the work to be open to the scrutiny of the press from the first. Many newspapers had already reported on many of the modifications Brewer trumpeted, outlining the changes the Curtiss team was forced to make in the restoration process for reasons of cost, safety, and expediency.

Nonetheless, for the next seven years, Brewer would repeatedly broadcast these modifications to the aerodrome in a determined effort to disparage Curtiss and the restoration project. In published writings and lectures on both sides of the Atlantic, he lambasted the restoration of Langley's aircraft as a premeditated hoax designed to blacken the Wrights' name.

Meanwhile, Orville was sufficiently exercised by Curtiss's activities that before the incident was over, he would also dispatch his older brother Lorin to Hammondsport to surreptitiously gather evidence of any changes the Curtiss team was making to the Langley plane to help discredit the undertaking in court. Armed with camera and binoculars, Lorin lurked around the hangars at Hammondsport, returning to stay in a hotel in the nearby town of Bath under the pseudonym of W. L. Oren. " I came here as I was afraid to telegraph from Hammondsport," he wired Orville during his visit. According to Lorin's account, he even had an altercation with Curtiss assistant Walter Johnson, who became suspicious of him and demanded that he either identify himself or hand over the film from his camera. Lorin reluctantly surrendered the film, but never divulged his identity or his mission.

•   •   •

The idea that Orville would send spies to the Curtiss camp seems at odds with the Wright brothers we normally remember—the earnest, young bicycle builders who attacked an age-old technological problem with fresh, ingenious thinking and dedication. And yet, secrecy and even spying are themes that echo throughout the Wright brothers' careers.

As soon as they got their powered airplane aloft on the winter sands of Kitty Hawk, North Carolina, in December 1903, their intense ambivalence about sharing word of their success is clear. On the eve of their famous flight, the telegram home to their father and sister says it all when the brothers write: "success assured keep quiet." As we now know, word in the press only broke about their initial flights when a subsequent private telegram home was, in the brothers' words, "dishonestly communicated to the newspapermen at the Norfolk office" and disseminated by the Associated Press.

If not for that leak to the press, in fact, it is unclear when the Wrights would have made their invention public. As it was, the timetable of the Wrights' activities speaks for itself. They flew successfully in December 1903. After a lengthy process, their key U.S. patent on their invention was granted on May 22, 1906. Yet, even then, the Wrights did not publicly demonstrate their airplane until the summer of 1908.

The timing says so much about the Wrights. They crossed one of the most momentous technological thresholds of all time—the centuries-old dream of skyward-looking inventors—but they showed their invention to practically no one *for four and a half years.* During that lengthy period, as the Wrights worked secretly on their airplane, they groped for how best to retain control of it. Like many inventors, they tried to garner the broadest ownership rights they could through the patenting process. Yet, even after their sweepingly broad patent was issued, the Wrights declined to publicly display their airplane for two more years while they worked actively behind the scenes to close licensing deals with many of the world's largest governments.

Enamored of the Wrights' technological achievements, most of their biographers are generous in accounting for these four and a half years of secrecy between Kitty Hawk and the Wrights' first public demonstrations of their airplane in France and for the U.S. Army in the summer of 1908. Some, for instance, citing the public backlash against Langley, argue that the press and the public were simply not receptive to news of the Wrights' accomplishment—undoubtedly a factor that may have contributed to their silence.

But no matter what the interpretation, it is safe to say that the Wrights' secrecy shaped—and retarded—the development of aviation in the United States and abroad. Of course, the Wrights had to work hard to keep such a big secret for so long. And in fact, there is compelling evidence that their secrecy was carefully calculated to maximize their control over their invention as well as their profit from it.

For instance, to perfect their design, the Wrights flew throughout 1904 on an open field at Simms Station outside Dayton. They would later claim in court that scores of passersby witnessed their machine in flight. But the statement is highly misleading. Throughout the earliest years of the airplane, the Wrights kept extraordinarily tight control over who they allowed to see their invention.

Presumably because they knew that witnesses might be helpful at some later point to attest to their accomplishments, the Wrights did allow a select handful of influential and discreet local friends and business associates to view their airplane. But, more notable are the great pains the Wrights took to deflect the attention of the press, even pretending their airplane wouldn't work when reporters showed up. Years after the fact, Wilbur alludes to their strategy in a letter to his brother while he was trying to drum up business in France: "No doubt an attempt will be made to spy upon us while we are making the trial flight," he wrote. "But we have already thought out a plan which we are certain will baffle such efforts as neatly as we fooled the newspapers during the two seasons we were experimenting at Simms."

Given the Wrights' obsession with keeping their airplane secret in its earliest years of existence, the first eyewitness press account of the Wrights in flight is also noteworthy. The article, a marvelously quirky description by Amos Root, appeared in January 1905 in an obscure newsletter called *Gleanings in Bee Culture*.

"I was right in front of it," Root recounts, "and I said then, and I believe still, it was one of the grandest sights, if not the grandest sight, of my life. Imagine a locomotive that has left its track, and is climbing up in the air right toward you—a locomotive without any wheels, we will say, but with white wings instead. . . ."

Despite the fact that many of the largest newspapers and magazines were keenly interested in the rumors circulating about the Wright brothers' invention, offering all sorts of deals for exclusive stories, the Wrights flatly turned these down. Somehow, though, Amos Root—ace reporter and beekeeper—managed to drive some two hundred miles on his own by automobile to Dayton and land the scoop of the century. The provenance of Root's reporting has never been proven but, as several historians suggest, it seems quite likely that his account was made at the clever invitation of the Wrights so that an early, independent published account of their airplane in flight would be available as proof of their ability to fly, but in such an obscure publication that it would never draw the attention of the press or the broader public.

The interpretation is given added credence in light of an incident that occurred in October 1905, when two trolleys happened to pass the prairie while Wilbur was airborne. Hearing the news, Luther Beard, managing editor of the *Dayton Journal,* ventured out to Simms Station to get the story. The Dayton paper did report the news on October 5 but, according to several accounts, the brothers bought out all the available copies they could. Then, with the help of several influential friends, they managed to suppress the story by keeping it from going out over the news wires.

The Wrights' approach, secretive and proprietary throughout, is even illustrated in their airplane itself. The brothers painted their early *Wright Flyer* gray so it would be harder for potential competitors—or "spies"—to photograph. In contrast, Curtiss and his crew went out of their way to gain public recognition. From the first, Curtiss and his team painted their aircraft bright colors. One of the first planes he flew, the *Gold Bug,* for instance was coated with yellow varnish expressly to help it show up more clearly for spectators and photographers alike.

As historian Robert Wohl notes in his work *A Passion for Wings: Aviation and the Western Imagination 1906-1918,* the Wrights "operated on the assumption that, if they sat tight and guarded their secrets, governments would eventually be forced to come to them and accept their terms." Throughout 1906, this account continues, "Wilbur believed that there was not one chance in a hundred that anyone would produce a machine 'of the least practicing usefulness' within the next five years." Unfortunately for the Wrights, they badly underestimated their competition, especially a team including Glenn Curtiss.

Perhaps even more sadly, the Wrights' proprietary strategy would take a personal toll, setting them at odds with some of their oldest and dearest colleagues. For instance, when the Wright brothers started out as a team of earnest young bicycle builders interested in the prospects of flight, they wrote to Samuel Langley at the Smithsonian, who gladly sent them reprints and citations of pertinent aeronautical research.

They also wrote to Octave Chanute.

Chanute, an eminent engineer of the period, who had made his reputation building railroad bridges throughout the country, was one of the nation's foremost experts on aviation in the late nineteenth century, a rare member of the scientific and engineering establishment at that time who was willing to devote himself in earnest to the then-heretical matter of human flight. Ultimately, Chanute would become a mentor to the Wright brothers and, like Langley, a central figure in the fledgling field of aviation. He corresponded widely and frequently with members of the small community and, in 1894, penned the influential *Progress in Flying Machines*—a book that greatly influenced the Wrights and many others seeking to unlock the mysteries of flight.

When they first solicited Chanute's help, in a letter dated May 13, 1900, Wilbur Wright had written: "I believe no financial profit will accrue to the inventor of the first flying machine, and that only those who are willing to give as well as to receive suggestions can hope to link their names with the honor of its discovery. The problem is too great for one man alone and unaided to solve in secret."

Wilbur's magnanimous remarks perfectly described Chanute's role among his early generation of aviators; Chanute corresponded so widely within the field that he served as a one-man clearinghouse, constantly linking practitioners up with one another, alerting them to new research, and even underwriting the efforts of some experimenters. In characteristic fashion, Chanute replied to the Wrights' letter immediately and gave freely of his accumulated wisdom about aeronautics. He supplied the Wrights with the latest technical literature, advised them each step of the way, and even helped them pick the location of Kitty Hawk, suggesting the mid-Atlantic coast for its steady winds and forgiving sand dunes.

But immediately following their first success at Kitty Hawk, Chanute began to notice a change in the Wrights' thinking. "We are giving no pictures nor descriptions of machine or methods at present," came the impersonal telegram in response to Chanute's eager inquiries about their experiments in December 1903.

By 1909, even Chanute, perhaps the Wrights' closest ally and mentor, broke off relations with them, charging publicly that their legal claims were overblown, greedy, and harmful to the nascent field. As Chanute explained it to a reporter from the *New York World:*

I admire the Wrights. I feel friendly toward them for the marvels they have achieved; but you can easily gauge how I feel concerning their attitude at present by the remark I made to Wilbur Wright recently. I told him I was sorry to see they were suing other experimenters and abstaining from entering the contests and competitions in which other men are brilliantly winning laurels. I told him that in my opinion they are wasting valuable time over lawsuits which they ought to concentrate in their work. Personally, I do not think that the courts will hold that the principle underlying the warping tips can be patented. . . . There is no question that the fundamental principle underlying [this] was well known before the Wrights incorporated it in their machine.

In biting letters early in 1910, Chanute was more pointed still. Even if they won in court, Chanute said, the strategy was a mistake. "I am afraid, my friend," Chanute wrote to Wilbur, "that your usually sound judgment has been warped by the desire for great wealth." The breach in their relationship would never be mended. In November of that year, at age seventy-eight, Chanute died unreconciled with the Wrights.

Orville's intransigence would become increasingly pronounced over the course of his life. The breach with Chanute would later be repeated in Orville's closest relationship of all—in a rift with his older sister Katharine. Throughout his life, Katharine had been a surrogate mother to Orville; their own mother had died when Orville was still a boy. Katharine had nurtured Wilbur and Orville's experiments at Kitty Hawk and helped in every facet of their work. She and Orville shared a house for the majority of their lives. But in 1926, at age fifty-two, Katharine fell in love and married a friend from her student days at Oberlin College, who had become an editor of a Kansas City newspaper. Feeling deserted and betrayed, Orville never forgave his sister for leaving him. After her marriage, he barely spoke to her again before she died just three years later in 1929.

In their dealings with others, even those closest to them, the Wright brothers—and Orville especially—could never be called magnanimous or generous of spirit. But, as the biblical saw goes, you reap what you sow. Orville's lengthy list of perceived wrongs and injustices painted him as a truly tragic figure toward the end of his life. In a detailed profile in 1930, a reporter from the *New Yorker* depicts him as "a gray man now, dressed in gray clothes. Not only have his hair and his moustache taken on that tone, but his curiously flat face. . . . [a] man whose misery at meeting you is obviously so keen that, in common decency you leave as soon as you can."

• • •

Personal animosities aside, the legal battle Orville Wright waged to uphold his exclusive control over the airplane in its first decade would also take on a life of its own. As Fred Howard, one of the Wrights' biographers, put it many years later, the bitter legal battle between Curtiss and the Wrights gathered size and momentum "like a large snowball rolled down a snowy hillside, leaving exposed in its wake . . . a sordid trail of hatred, invective, and lies that muddy the pages of aeronautical history to this day."

Ultimately, the case would cripple the development of the youthful aviation industry, especially in the United States. The effects are obvious in retrospect. The field was torn into rival factions. Experimentation was discouraged; investment tied up and, most notably, as the Wrights waged a total of nearly three dozen lawsuits, the legal wrangling siphoned energy away from building airplanes and tangled the field in knots of accusations and uncertainty. Many of these effects were felt at the time. As an editorial in the *Boston Transcript* newspaper put it after Orville Wright's assertion of "absolute control" over the industry: "The effect of the Wright decree will beyond question numb what little life remains today in aviation in America."

Aviators, in particular, resented the bitter and litigious climate that had engulfed their fabulous new flying machines. Charlie Hamilton, a student of Curtiss's who would earn a place in the history books for his flights of unprecedented duration, quipped that the field had added a new prerequisite. As he put it, "A man has to have ten years in law school before he has a chance of becoming an aviator. "

Consequently, by 1914, anti-Wright sentiment ran high—not only in Hammondsport. Curtiss's workers, like most aviators around the world at the time, resented the legal action the Wrights had instigated and saw Orville as a spoiler. Even *Aeronautics,* one of the most widely read periodicals in the field, ran a lengthy article by a prominent patent attorney analyzing the basis of the Wright lawsuit and claiming that their case was overblown. A close reading of the Wright claims, the lawyer asserted, showed that they "do not cover supplementary surfaces," namely the separate, adjustable wing flaps called ailerons. The interpretation surely influenced the attitude of flyers toward the case even if it did not ultimately sway the courts.

As another outspoken editorial noted, Orville Wright "will never suffer for want of this world's goods. His name and fame will suffer, however, if, instead of contributing his future interest and enthusiasm to the further conquest of the air, he sulks in his tent and blocks the game of other people on account of his parsimonious concern for an old patent."

With the fate of an emerging industry literally in the balance, and with the prospect of a high-stakes spectacle, the aerodrome case soon drew front-page headlines in the press, moving to center stage in the bitter, ongoing drama over the control of the emerging aviation industry. The appearance of a conflict of interest on the part of Curtiss and the Smithsonian team, not to mention the existence of the bitter lawsuit with Orville Wright, soon led to a full-blown imbroglio.

Unfortunately, the posturing and name calling obscured an underlying issue of historical significance. After all, at its heart, the aerodrome restoration raised profound issues of precedence and posterity. About how technological change occurs and how history remembers it. As a *New York Times* editorial observed, the Wright suit "was won upon the fact that no other aeroplane had ever maintained itself in air with human freight, and inferentially could not. What effect Mr. Curtiss's aerodrome restoration project might have in modifying the recent decision of the circuit Court in favor of the Wrights' connection no one can now tell."

THREE

AMERICA OR BUST

*The Atlantic Ocean, one of these days, will be no more
difficult to cross by air than a fish pond.*
—GLENN H. CURTISS

On a warm spring evening in 1914,
several days after the arrival of Langley's aerodrome, Glenn Curtiss
waits on the open-air platform of the Hammondsport station beside
scenic Lake Keuka. From his perch there at the edge of town, he can
see the familiar and sparsely populated wooded hills that slope
down to the Lake's distant bank. All is quiet. He hears only the
murmur of crickets and occasional wafting voices from far across
the water until the raucous clatter and roar of the approaching train
rises to drown them out.

As the train shatters the evening's tranquillity, so will its arriving
passenger presage jarring changes for Hammondsport in the spring
and summer to come. On board is Albert Zahm, a dapper man in a
carefully pressed suit and straw boater hat. Zahm, the liaison from

the Smithsonian Institution, is coming to town to oversee the particulars of the aerodrome's reconstruction.

Zahm's arrival will mark the start of a swirl of activity unlike anything the residents of Hammondsport have ever seen. Aviation luminary Charles Manly, Langley's former assistant, will soon join the restoration effort as will Charles Walcott, head of the Smithsonian Institution. Many others with no connection to the aerodrome project will descend upon Hammondsport as well. Elmer Sperry Jr., for instance, comes to town to perfect the automatic airplane stabilizer Glenn Curtiss helped him develop. Earlier in the year, Sperry stunned crowds in Paris with the invention, as he left his airplane's controls in mid-flight over the Seine while his mechanic climbed out onto its wing. Like many others interested in aviation, Sperry chooses Hammondsport as the best place to advance his aeronautical work.

In the months to come, interest in the Curtiss Aeroplane Company will reach new heights. Curtiss is hamstrung by his lawsuit with the Wrights, but he has managed to continue nonetheless, with orders for Curtiss airplanes and so-called flying boats—the single-hulled seaplanes that Curtiss had invented—arriving from as far away as India, Russia, and Japan. At home, flying boats will increasingly draw the attention of well-heeled executives. Henry Ford makes the pilgrimage to Hammondsport to marvel at Curtiss's latest invention. Harold F. McCormick, vice president of the International Harvester Corporation, comes to purchase a flying boat on the spot. He says he wants to use it for the commute from his Lake Michigan estate to his office in Chicago. And activity at Curtiss's flying school will boom as never before with eager new pilots venturing out each day, weather permitting, to practice over Hammondsport's Lake Keuka.

As Zahm steps from the train, Curtiss spots him at once and

approaches to greet him warmly. But there is, on both sides of their formal handshake, a mixture of admiration and apprehension. No amount of cordiality can overshadow the dramatic differences between them. Curtiss has spent his whole career learning by doing. He has had no formal schooling in aeronautics and does not easily translate his remarkable grasp of aviation principles into words. Zahm, a distinguished professor of aeronautics, has written the field's most widely read text. Yet Zahm has had remarkably little firsthand experience with aircraft.

The two men differ in temperament as well as training. Curtiss is intuitive and spontaneous; Zahm is formal and methodical. And, of greater significance to the restoration, Curtiss, swamped with the demands of a burgeoning business, has thought little about the fine-grained details of the restoration project; Zahm has, for more than a month, thought of little else.

Bridging this divide is not easy. But after a formal, even awkward, exchange upon Zahm's arrival, he and Curtiss will build an enduring friendship. In addition to their mutual love of aviation, the fact is that each comes to hold the complementary talents of the other in particularly high regard. Zahm writes much later that, from the first, he marvels at Curtiss's natural way with his associates and workers, calling it a gift that manages somehow to inspire their most "enthusiastic efforts."

The buttoned-up Zahm marvels too at Curtiss's unnervingly casual manner about his dog Terence—who Zahm describes as "a Scotch roustabout." Beloved throughout Hammondsport, Terence enjoys the free run of town and has no greater joy than jumping excitedly into Curtiss's automobile or those of the local champagne merchants to go for a ride.

In the months to come, Zahm will help guide Curtiss through a reconstruction effort that could well alter aviation history. For his part, Curtiss will draw Zahm out of the ivory tower and fully into the tumultuous, emerging aviation business. That summer, with Curtiss's encouragement, Zahm will even become a pilot. After his first outing with Curtiss's lead flight instructor Francis Wildman, Zahm calls it a thrill to "couple the practical with the theoretical" after studying the mechanics of flight for so many years.

The morning after Zahm's arrival, most likely over breakfast at Curtiss's home, the two begin to hammer out the particulars of the aerodrome restoration.

When Charles Walcott, secretary of the Smithsonian Institution, first formally broached the possibility of restoring Langley's aerodrome, Curtiss promptly sought the opinion of Alexander Graham Bell. "I should like to know what you think of this plan," Curtiss wrote to Bell in February, "as it would be an easy thing to do provided it is worthwhile."

But while Bell and others agreed the project had merit, Curtiss is now coming to realize that the task—fraught as it is with serious historical weight—will not prove so easy after all. As he listens to Zahm's detailed concerns, Curtiss confronts the tangled complexities involved in trying to remain unimpeachably faithful to Langley's original design.

The first thorny problem is the aerodrome's method of takeoff. Curtiss says he is prepared to try a newly designed catapult like the one Langley employed. But, after some discussion, both he and Zahm agree that such a scheme is too dangerous, untested, and costly. The assessment leaves two alternatives: trying to launch the aerodrome from land, or from water.

Given that Curtiss recently won the Langley Medal for the hydro-aeroplane, not to mention his frequent reliance on Lake Keuka for testing his new designs, it is not surprising that Curtiss favors a water launch. His plan is to attach pontoons to the bottom of Langley's otherwise unchanged design and see if they can get it to rise from the lake's surface on its own power.

Curtiss makes it sound like a simple, straightforward procedure, but the addition of pontoons will mark a dramatic change for the aerodrome. First of all, pontoons are heavy. The 1903 aerodrome is not long on thrust to begin with and the pontoons will inevitably add hundreds of pounds to the aircraft as well as increase its wind resistance. In addition, retrofitting pontoons to the aerodrome will almost certainly require additional bracing to accommodate the added weight and air pressure. Will the changes they contemplate preclude an accurate test of Langley's aircraft?

Decisions like the one to add pontoons pull the project deeper and deeper into uncharted terrain. As the work progresses, critics, including a near-livid Orville Wright, will as much as accuse Curtiss and Zahm of conspiracy and fraud: making secret changes to the aerodrome to improve its chances of flying. But the shrill charges do not bear up under close scrutiny.

As Zahm later recounts, he and Curtiss agree upon a two-phase set of experiments. In the first phase, their goal is to undertake a meticulous restoration with as few changes as possible to the original aircraft to see whether the aerodrome, as originally constructed, is capable of sustained free flight with a pilot. In a second, separate effort, they contemplate making some significant changes—including the addition of a modern Curtiss engine—to more comprehensively test the aerodynamics of Langley's tandem-wing design. Many of the charges that will soon be leveled by

Orville and Brewer will conflate these two clearly separate goals.

Curtiss and Zahm do authorize changes in even the initial reconstruction phase. But most of these, as Curtiss will put it later, do
nothing more than ensure that Langley's machine will be "not quite
as good as new" by the time they are finished. Most of the changes
they sanction, like the pontoons, worsen the plane's prospects to
get aloft. For example, the aged motor, producing only 40 of its
original 52 horsepower, cannot turn the propellers at the requisite
950 revolutions per minute stipulated by Langley. To compensate
for this problem, Curtiss and Zahm opt to trim the propeller blades
to allow them to spin faster. Critics will rightly charge that the
tapered ends they create alter Langley's precise design and take
advantage of aeronautical knowledge Langley never possessed. But,
such quibbles aside, even the critics reluctantly acknowledge that,
in the end, the rebuilt aircraft will have less thrust than the ones
Langley originally designed.

Throughout, Curtiss and Zahm argue strenuously that the
changes they authorize do nothing to hinder their ability to fairly
test the aerodrome. As Zahm will note, getting the aerodrome to fly
with pontoons is like trying to get an eagle to fly "with a child in its
talons." Should the eagle fly under such circumstances, he explains
tersely, a scientist ought to be able to accurately infer that the bird
could fly at least as well unfettered.

Nonetheless, given the highly polarized state of the emerging aviation industry, it is not surprising that any change from Langley's
exact design provides fodder for controversy and dissension. As
scrupulous as Curtiss, Zahm, and the rest of the team may try to be,
nothing can refute the Wright camp's charges of a fundamental conflict of interest. After all, Curtiss, longtime adversary of the Wright
brothers in court, is now helping to restore an aircraft that could

cast doubt on the Wrights' standing as the first with a viable, piloted airplane.

For his part, Smithsonian Secretary Walcott contends that his and Zahm's supervision of the effort ensures that it is an unbiased scientific experiment. Of course, Walcott certainly realized that Curtiss was involved in a lengthy patent dispute with the Wrights when he chose Curtiss to oversee the work. But, at least at the outset of the reconstruction process, he didn't view *Wright v. Curtiss* as a problem. As Walcott jauntily tells the *New York Herald,* he "never gave the patent situation a thought" when he commissioned Curtiss to restore the Langley machine.

Of course, Walcott's long association with Langley can be seen to undercut his claims of neutrality as well. Not only is he Langley's successor at the Smithsonian but, in his former position at the U.S. Coast and Geological Survey, he played a key role in persuading President McKinley to appoint the board that authorized Langley to build the full-scale aerodrome in the first place. A successful restoration project will thus, in some sense, vindicate him as well as Langley. Still, it is one thing to have allegiances, quite another to engage in outright fraud. And there is no evidence of fraud or conspiracy on the part of Walcott or any member of the distinguished and reputable team engaged in the restoration project.

Nonetheless, once the restoration work is under way, Walcott and the rest of the team will contend with cries of foul from the Wright camp, including many accusations that have never been fully settled a century later. In fact, long years of acrimony so cloud the issue of the aerodrome's reconstruction it is difficult to strip away the charges and countercharges sufficiently to gain a clear picture of what transpired.

Thanks to the intense interest of the press in the matter, however, it is clear that charges of secrecy in the restoration process are unfounded. Even a cursory look shows that newspapers carried reports on the aerodrome restoration as it went along, replete with detailed descriptions of the alterations Curtiss and Zahm authorized as the process moved forward. It is to their credit that Curtiss and Zahm kept the work open to the press and other interested parties. But, when they embarked on the project, they surely had no idea about the amount of interest the story would ultimately generate.

In the spring and summer of 1914, the reconstruction of the Langley aerodrome is big news and the daily happenings in Hammondsport suddenly make their way to the front pages of newspapers everywhere. Reporters descend upon the town from as far away as London. Some, like J. Clarke of the *New York Sun,* Herbert Swope of the *New York World,* and Joe Toy of the *Boston American,* file almost daily updates on the progress of the aerodrome. The interest of these members of the press—well-traveled representatives from the bustling world beyond the Finger Lakes—lends a sense of interest and self-conscious pride that pervades the town.

By 1914, with the breathtaking antics of stunt fliers like Link Beachey and a series of air meets and exhibitions around the country, the American public is enthralled with the airplane. And now Curtiss, with the blessing of the Smithsonian Institution, is undertaking nothing less than to try to rewrite its brief history.

As the warm spring days give way to summer, there are easily enough journalists in Hammondsport to field a baseball team. And as they await the latest news from the Curtiss factory, they even take advantage of the fact, playing several spirited ball games against the aviators in a nearby cow pasture before scores of local spectators.

After all, like airplanes, "America's pastime" is young and popular; that summer, in fact, a young rookie named Babe Ruth has just signed with the American League.

In Hammondsport, the Curtiss team boasts some competent hitters, but the journalists are the stronger team overall, even though their talent is wildly uneven. In one incident, highlighted in the local press, Charlie Stiles of the *New York Tribune* is so startled when he finally hits the ball he forgets to run to first base.

As noteworthy as the aerodrome reconstruction is, the reporters do not remain in Hammondsport for that story alone. The fact is, Curtiss makes good copy. Even though Orville Wright has done everything in his power to shut down Curtiss's operation, the reporters marvel that Curtiss somehow manages to remain unflaggingly optimistic and committed to his pursuit of aeronautical engineering.

In the spring of 1914, Curtiss vows to take his case against Orville Wright to the U.S. Supreme Court. But it is hard to see what legal grounds he might use to convince the high court to actually hear the case. Posturing aside, Curtiss stands in an extraordinarily precarious position. With few remaining options, the prospect of caving in to Orville Wright could mean his company's demise. As J. Clarke explains in the *New York Sun:* "The patent situation gives Orville Wright a practical monopoly of the aeroplane industry in the United States and no manufacturer has yet been given a license." In Curtiss's case, Clarke reports, "the back royalties demanded by Wright would amount to more than the capital stock of the Curtiss Company and the Wright Company taken together, according to a statement of Orville Wright to the *Sun* representative in New York recently."

What will Curtiss do? The looming questions about the viability of his business only heighten the interest of the press. There always seems to be something new to report. One option that causes a good deal of speculation is that Curtiss might move his operation to another country to get out from under the cloud of the Wrights' overly broad U.S. patent. According to Curtiss's publicist Lyman Seely, Curtiss receives offers from at least three European countries. One foreign government, Seely tells reporters, has promised that if Curtiss relocates, it will guarantee to double what his firm earned in its best year at Hammondsport.

Meanwhile, in an effort to avoid infringing the Wrights' patent, Curtiss is now producing airplanes with "nonsimultaneous ailerons." Unlike the Wrights' patented system, the flap on each wing operates independently. It is not an optimal system, but at least it allows Curtiss to continue to manufacture for the moment. Orville Wright has already threatened to haul him back into court over the matter. Consequently, that spring Curtiss begins to investigate the possibility of making his airplanes' ailerons in Canada, going so far as to dispatch a member of his staff to price commercial real estate across the border.

Despite his seemingly insurmountable legal difficulties, Curtiss is aware that a key variable has changed. He has become the favored underdog and the tide of public opinion has swung to his side. One newspaper editorial from as far away as Jackson, Mississippi, sums it up that spring, noting that "the Almighty Dollar was looming up in the vision of Mr. Wright and his associates to the exclusion of all regard for the future of aviation or the good of humanity."

With such public sentiment, after years of little financial security, Curtiss now finds that he has some powerful supporters of his own. Foremost among these is W. Benton Crisp, the most famous patent lawyer of the day, and the press notes with excitement his arrival in Hammondsport that summer. Crisp, who heads Henry Ford's legal staff, has just won a bitter and much celebrated case against a man named George Selden who claimed to have been the first to patent the motorcar. In fact, Selden, a lawyer and part-time inventor, did hold the first automotive patent but, somewhat like the Wrights' claim, it was extremely vague and broad. Nor did Selden ever actually manufacture any automobiles. He chose instead to try to extract money from carmakers like Ford. The case drove Henry Ford to distraction; he was firm in the belief that patent protection should be used to bring new innovations to market, not to stifle competition.

Identifying with Curtiss's problems, Ford has stunned the public by magnanimously offering Curtiss his winning legal team and any other help he might need in his case against Orville Wright. The reporters know Crisp's arrival in Hammondsport signals that the case may still have some life in it after all.

For his part, Curtiss doesn't seem outwardly discouraged by the dire state of the Wright lawsuit. Much to the endless fascination of the press, he even seems to be enjoying himself. There are few days off for Curtiss in the spring and summer of 1914. When he does take a break, the reporters often find him tinkering on some new idea or putting on some kind of show for their benefit. One hot day during this period, for instance, Curtiss rigs an engine to a pontoon to create a water craft he dubs a "sea sled," precursor to a modern-day jet ski. The machine works moderately well but the major result, much to everyone's amusement, is to land Curtiss repeatedly in the water.

Each day seems to bring new twists to Curtiss's story. And yet, he appears to be at the top of his game. At the age of thirty-six, Curtiss has seen something of the world, and over the spring and summer of 1914, he uses all his talents to oversee a multi-ring aviation circus, upping the ante by adding flaming torches to his routine just when the audience thinks he has already reached his limit.

That spring, taking advantage of the favorable attention from the press, Curtiss announces his new show-stopping venture: a plan to build a seaplane capable of crossing the Atlantic Ocean.

Underwritten by Rodman Wannamaker, an American airplane enthusiast and heir to a department store fortune, Curtiss's proposal seems so outlandish that Orville Wright confidently tells the press that Curtiss will surely fail. After all, the proposal comes some thirteen years before Lindbergh's historic nonstop crossing.

What would Orville say, reporters ask, if Curtiss were to succeed?

"I have not enough expectation that the craft will ever land near enough to any country where our patents are valid—that is, anywhere in Europe—to make it worthwhile to tell you what I would do in that case," Orville snips. According to Orville, the whole idea is nothing more than a publicity stunt. Curtiss, he says, likely knows full well his machine will not be able to cross the ocean.

Even Curtiss might admit that the crossing is a long shot, but he cannot tolerate it when Wright impugns his integrity. As Curtiss quickly retorts, "The question as to whether or not I am sincere in this undertaking is answered by the fact that payment for the machine is due only after it has proved itself capable of carrying the load sufficient for the proposed flight. It is obvious that I would not undertake this arrangement unless I was confident of being able to produce the machine."

Further, Curtiss adds testily, "I call attention to the fact that there has been considerable advancement in aviation since the Wrights made their machine. There are many practical engineers, aviators and scientists who believe that the Atlantic will be crossed by aeroplane this year."

The fact is, unknown to Orville, Curtiss had been quietly thinking of crossing the Atlantic for at least two years, ever since he perfected the flying boat. In 1912, Curtiss even discussed the matter with Lieutenant John Towers of the U.S. Navy and several other members of the U.S. Aero Club, noting that he was so encouraged by the success of his hydro-aeroplane, he had begun to take "great interest in the idea of a flight across the ocean by aeroplane."

The plan took one step closer to reality in the spring of 1913 when Lord Northcliffe, publisher of London's *Daily Mail* newspaper, offered a prize of $50,000 for the first transatlantic crossing in either direction by a hydro-aeroplane. According to the ground rules set forth for the competition, the vessel could touch down along the way if necessary, but would have to complete the crossing in seventy-two hours or less.

For Curtiss, the dream of an Atlantic crossing started to actually take shape early in 1914, when Wannamaker agreed to sponsor Curtiss to build the aircraft. With Henry Ford's backing, the notion became a full-blown plan.

Orville Wright, despite his disparaging comments, was taking no chances. Apparently, Wright believed enough in the project to threaten to halt the effort with a court order as another alleged infringement of his now-obsolete wing-warping patent.

Luckily for Curtiss, Orville's threat does little to scare off the project's backers. After all, Wannamaker argues, the crossing will be an entrant in Lord Northcliffe's contest; Curtiss will not be manufacturing the airplane for sale nor, strictly speaking, for exhibition proceeds and therefore cannot be accused of infringing any patent. Plus Wannamaker argues, his deal with Curtiss was negotiated in

February, prior to the court's most recent ruling in favor of the Wright patent.

Orville must find these arguments persuasive and, in this case at least, backs off from his threat to shut the project down.

Activities in Hammondsport reach a dizzying crescendo of excitement on June 22, 1914, as the transatlantic aircraft—called the *America*-—makes its public debut. On a sunny Monday afternoon, a crowd of nearly two thousand flocks to the banks of Lake Keuka to witness one of the world's great wonders. Automobiles filled with spectators arrive from far and near. Men in boaters and summer-weight suits, and women in long dresses gather before a stage erected in the park by the water's edge, while local boys in white shirts and knickers squeeze their way to the front for a closer look. All eyes are riveted on the sleek and enormous flying boat that promises to make the momentous ocean crossing. No fewer than three separate crews bring movie cameras to chronicle the event.

It is certainly unlike any airplane unveiled before. An enormous biplane, *America* has two sets of wings atop one another that stretch an unprecedented 74 feet across. Between the wings are fitted not one but two 100-horsepower Curtiss OX motors—the largest the company makes. Like many airplanes of the day, it is a so-called "pusher" aircraft: the motors drive two big propellers that sit behind the wings to push the craft forward.

Centered below the vast wings is a single fuselage enclosed in deep red, laminated silk with a V-shaped hull like that of a speed-boat. And to face the rigors of the inclement crossing, the cockpit is fully enclosed, another virtually unprecedented development in air-plane design. Sitting atop the boatlike hull, the cockpit looks like a

compact version of the wheelhouse on a tugboat, with a curving cel-
luloid windshield to afford the craft's two pilots a sweeping
panorama of the world below.

The crowd presses as close as it can before the large, temporary
platform erected on the edge of Lake Keuka, next to the lakeside
spot where the *America* has been hauled. Glenn Curtiss stands confidently upon the stage, flanked by the two men he has tapped to
pilot the flight, Lieutenant John Cyril Porte of the British Royal
Navy, and George Hallet, one of Curtiss's most trusted mechanics.
With a nod to the British sponsors of the event, Curtiss has chosen
Porte for his navigational expertise. Hallet, meanwhile, makes up
for his relative lack of flying experience with an intimate knowledge
of *America*'s mechanical systems. Presumably, if need be, he can
make crucial adjustments to the aircraft en route. Hallet has even
practiced changing spark plugs in mid-flight.

Also on stage is Katherine Masson, sixteen-year-old daughter of
a well-known local vintner, who has been selected by lot to christen
the aircraft. Her summer bonnet is set fashionably askew as she
strives for poise before the unnervingly large crowd.

As Curtiss has recently announced to the press, the initial plan is
to make the trip in three hops, flying from St. John's, Newfoundland, to the Azores Islands off the coast of Spain, a distance of some
1,200 miles, and from there to the coast of Spain, 600 miles, before
embarking on the final 500-mile leg to England. Curtiss's goal is to
begin a quick round of tests and ready the plane for a July crossing
to begin, weather permitting, when the moon is full.

After Miss Masson recites a poem composed for the occasion by
none other than Dr. Zahm of the Smithsonian Institution, it is time
for her to officially christen the aircraft by breaking a bottle of
locally produced Great Western champagne across a scaffolding

erected next to *America*'s fabric-covered prow. But despite repeated efforts, she can't seem to get the bottle to break. Lieutenant Porte comes to her aid with a little more force. But even he can't manage to smash the bottle, leading Curtiss to intervene out of concern for the aircraft's delicate hull. With the crowd in an uproar of laughter, Porte resorts to crushing the bottle with a handy sledgehammer, bathing himself and the others on stage in an explosion of spurting champagne.

Then, with the crowd still boisterous and elated, a team of forty men launches *America* into the lake. Despite the enthusiasm, however, it is too late in the afternoon to do much more and the festivities gradually draw to a close for the day.

After hours more of mechanical work assembling the machine and checking its fittings, *America* will have her maiden voyage the following day around 3 P.M., amid the cheers of the returning throng. Curtiss and Porte climb aboard and take a short spin. The craft skims the surface of the water, never lifting far from Lake Keuka, but the two pilots return beaming, obviously pleased with the initial trial. "She is as strong as a blooming rock," Porte offers. "She has tons of horsepower and is very stable."

"I am more than satisfied," Curtiss adds. "We did more than I intended to do today. The boat has come up to my best expectation and there is no reason why she should not do the work she is built for."

Before the week is out there is no doubt that the *America* is causing an unprecedented stir. Locally, photographer Hank Benner reports having sold an astonishing seventy thousand postcards of the *America* and, facing strong demand, plans to print more. Internationally, the plane is making waves as well. The eminent banking firm Lloyds of London, which originally posted betting odds at 47 to 1 against the success of Curtiss's aircraft, revises its estimate more than sevenfold in Curtiss's favor. The new odds of 6 to 1 against the project's success do not exactly represent confidence in the *America,* but they do indicate the project's apparent momentum. The transatlantic flight looks more practicable with each passing day.

As the pieces come together, Curtiss's employees are gaining confidence as well. The 100-horsepower motors, for instance, prove themselves in the factory by running flawlessly nonstop for 100 hours. In fact, many of the workers are chagrined with the change at Lloyds of London. When the steep odds were posted, a group of workers at the Curtiss plant collected a pool of $2,000 to wager on *America*'s success. But by the time they cabled their bet, the enticing windfall had substantially diminished. Now if the flight proves successful, they will have to make do with a sixfold return on their investment.

Despite their confidence, much remains to be done and Curtiss and his workers badly underestimate the complications ahead. In fact, Miss Masson's ordeal with the champagne bottle proves a portent of the difficulties involved in getting *America* aloft.

In what will become a protracted source of frustration, preliminary load tests prove disappointing, forcing Curtiss to delay the target date for takeoff. The team comes tantalizingly close to their goal but cannot seem to reach it. The *America* needs to lift 5,000 pounds to carry sufficient fuel for the voyage. Yet, after a series of efforts, they can get only 4,607 pounds of ballast aloft.

Try as they might, the Curtiss team cannot find the additional thrust needed to lift those extra 400 pounds. By Zahm's count, Curtiss will make some twenty-eight separate changes in *America*'s design over the ensuing weeks in a determined effort to solve the problem.

The two OX engines just can't seem to do the job. Improvising, Curtiss decides to add a third, centrally mounted engine for use only during takeoff. But the elusive trick is to balance weight and power. Not only does the third engine add weight to the aircraft; its propeller windmills when it is idle, causing so much drag the plane burns enough extra gas to throw off the team's careful calculations about fuel allotment.

During these frantic weeks, Curtiss even borrows from other ongoing projects. He temporarily fastens pontoons designed for the Langley aerodrome onto the *America* in the hopes that the additional fins might help the plane to lift its load.

The work is often frustrating, but in characteristic fashion, Curtiss—part captain of the bold endeavor, part host of the ongoing gala—takes pains to ensure that the interest of the press and spectators doesn't flag. He consistently makes time to answer the reporters' endless questions and even brings ten reporters on board *America* for a brief ride around the lake to give them something new to write about.

One day, when things are looking particularly discouraging, he takes the *America,* weighed down by 5,000 pounds of sandbags, for yet another trial run on Lake Keuka. The aircraft still has difficulty rising from the water and skims across the lake. Frustrated, Curtiss waits until the plane is far enough out on the lake to be out of sight of spectators on the shore. Then, with help from co-pilot Hallet, he surreptitiously jettisons enough sandbags into the lake to allow the plane to come soaring back triumphantly toward the astounded spectators.

Curtiss owns up to the mischief, but not until after he basks in the crowd's cheering disbelief at the apparent success of the trial. The team, he says, just needed a reminder that success is close at hand.

If he worried about losing his audience, though, he shouldn't have. There is simply too much aviation history to be written in 1914. And the now-resident reporters have become fixated on learning all they can about the audacious and irrepressible Glenn Curtiss: about his perseverance and his unflappable nature; about how his dispute with the Wrights ever became so bitter; and about how, despite it all, this still seems to be his finest hour.

By June of 1914, powerful international forces are at work. The very same front page of the *Hammondsport Herald* that carries banner news of *America*'s christening bears an ominous and fateful headline from Europe: "Archduke murdered. Heir to Austrian throne and wife killed."

With hindsight, of course, we know that World War I is imminent. But, in bustling, rural Hammondsport in the spring and summer of 1914, all eyes remain riveted on Curtiss: Can he get Langley's ungainly aerodrome aloft? Can he possibly get the *America* to cross the Atlantic? And will he manage to keep building aircraft in the face of Orville Wright's tightening monopoly control of the emerging aviation industry? His prospects, in each case, seem slim. But the longer the reporters in Hammondsport watch Curtiss in action, the less willing any of them are to rule him out. And they are not about to leave town without seeing these stories through.

PART II

REACHING FOR
THE SKY

FOUR

CAPTAINS OF THE AIR

To fly is everything.
—OTTO LILIENTHAL,
GERMAN AVIATION PIONEER, 1890

Glenn Curtiss first met Orville and Wilbur Wright in Dayton, Ohio, on a blustery September day in 1906. Since Curtiss and the Wright brothers were blown together by the winds of history, perhaps it is fitting that a chance gust actually first brought them face-to-face.

Curtiss, then just twenty-eight years old, had come to town to accompany the world-famous "Captain" Thomas Baldwin at the Dayton Fair where, in a top-billed and much-anticipated event, Baldwin was to demonstrate his new *City of Los Angeles* airship. The airship was a dirigible, one of only a handful in the world in 1906. In a few short years, of course, dirigibles would be eclipsed by the airplane. But for the moment, they commanded center stage. Ever since the Brazilian aviator Alberto Santos-Dumont flew one around the Eiffel Tower in 1901, these odd aircraft had become the subject of intense public curiosity and amazement.

Unlike hot-air balloons—formerly the only reliable known method for getting aloft—dirigibles like Baldwin's were big oblong gasbags filled with hydrogen. To fill the bags, the intrepid fliers would pour sulfuric acid on a barrelful of iron filings and then capture the vaporous result, taking pains not to ignite the volatile gas in the process. Perhaps most notably, these aircraft were dirigible—or steerable—thanks to the inclusion of one or more propellers driven by a lightweight motor normally housed on one end of the pilot's catwalk beneath the hydrogen-filled bag.

Baldwin's dirigible was powered by an engine specially designed by Curtiss, who was already gaining acclaim as one of the world's preeminent designers of lightweight gasoline-powered engines. Curtiss welcomed the opportunity when Baldwin offered him a week's salary to tend to the airship's motor between its demonstration flights in Dayton. Thanks to Baldwin, he was beginning to find a whole new airborne market for his engines.

In fact, Curtiss had agreed to come to Dayton at least partly for the opportunity to pay a call on the Wright brothers, the mysterious bicycle builders who were rumored to have successfully gotten aloft in a heavier-than-air flying machine. Curtiss's interest was straightforward enough: he was looking for business. Baldwin seemed well pleased with his motors; Curtiss hoped the Wrights might be interested in purchasing them as well. Curtiss had even made this suggestion to the Wrights in a letter he sent prior to his trip to Dayton. As of his departure, he had received no reply.

By this time, the Wrights have long since undertaken their successful experiments at Kitty Hawk, North Carolina, and have continued their aeronautical work at Simms Station near Dayton. For more than a year, however, they have suspended their experiments altogether to try to secure lucrative contracts for their flying machine with governments around the world. The Wrights' plan is badly hampered by their secrecy. Even with the issuance of their broad "wing-warping" patent in the United States earlier in 1906, the Wrights have resolved to keep their airplane locked up and have yet to display it publicly. Although they guarantee their plane will fly, they insist that even the government officials they approach as prospective buyers must agree to purchase their technology sight unseen. It is a stiff requirement coming from two little-known bicycle builders with no advanced engineering degrees or prior record of invention. Except for a $5,000 deposit in a failed negotiation with the government of France, the strategy has yet to earn the Wrights any money.

Their behavior has also engendered a good deal of skepticism in the fledgling aviation community. Many are persuaded that if the Wrights' invention really worked, the brothers would have already demonstrated it, or at least the press would have ferreted out the news. As the editors of *Scientific American* note early in 1906:

Unfortunately the Wright brothers are hardly disposed to publish any substantiation or to make public experiments, for reasons best known to themselves. If such sensational and tremendously important experiments are being conducted in a not very remote part of the country on a subject in which almost everybody feels the most profound interest, is it possible to believe that the enterprising American reporter, who, it is well known, comes down the chimney when the door is locked. . . . would not have ascertained all about them and published them long ago?

Considering the Wrights' penchant for secrecy, it is not surprising that they disdained the aviation exhibitions that have been hosted with increasing frequency in the earliest years of the twentieth century. Not only do they shun these events as participants; they normally choose not to attend them as observers, either. But they will make an exception for Baldwin's demonstration. After all, in 1906, Baldwin is arguably the most famous outdoor attraction in the world. In this seminal period in the earliest days of aviation, he has performed death-defying aerial feats around the world—in North America, Europe, and the Far East. Whether ascending in one of his airships or jumping from impossible heights with a parachute, "Captain Baldwin's name," as one newspaper report put it, is "always a promise of thrills."

On the first day of Baldwin's visit, as the crowd begins to gather at the fairgrounds for his air show, the weather looks increasingly ominous. The wind has picked up enough that spectators are bracing themselves against it; men are quite literally holding on to their hats while women in the crowd attend their billowing skirts. It looks to Curtiss far too windy to attempt a flight. But the portly Baldwin, looking much like a circus ringmaster in tall boots, a dramatic dark cape, and bowler hat, seems to be measuring the gusty wind against the size of the growing crowd. He is torn. The wind will make it nearly impossible to control the dirigible but he has surely taken greater risks in the past. He decides to make the ascent despite the wind and clambers into the pilot's catwalk to start the dirigible's engine.

Curtiss has yet to recognize them but, sure enough, the Wright brothers have moved prominently to the front of the crowd to watch Baldwin's flight. They stand straight-backed, and side by side, decked out, as always, in neatly starched collars and fully buttoned suit jackets.

Glenn Hammond Curtiss at the controls of one of his early airplanes—
most likely the 1910 *Hudson Flier*.

Charles Manly *(left)* and Samuel Pierpont Langley *(right)* stand atop the houseboat they used to try to launch Langley's pioneering "aerodrome" in 1903. The iron catapult they used can be seen in the background, as can the compass Manly fastened to his left trouser leg.

The sole known remaining photo, from the *Washington Star*, captures the aerodrome's crash on December 8, 1903. The houseboat launching pad can be seen in the right foreground, as can the aerodrome's twisted, crumpled "Penaud" tail.

Glenn Curtiss's early bicycle shop in downtown Hammondsport , New York, circa 1901.

A bird's-eye-view rendering of the Curtiss Manufacturing Plant in Hammondsport in 1907. This drawing was included in catalogs Curtiss used for selling his motorcycles to customers around the country.

The inside of the Curtiss shop, circa 1907. Early models of Curtiss's two-cylinder motorcycles can be seen in the foreground. A model very similar to these caught the eye of aviation daredevil Thomas Baldwin, and gave him the idea to use a Curtiss engine to power his early dirigible.

A determined-looking Curtiss, winner of numerous motorcycle races, at the wheel of one of his early models. Careening at 136 miles per hour on an even larger, eight-cylinder motorcycle of his own design in Ormond Beach, Florida, in 1904, Curtiss would earn himself the title "the fastest man on earth."

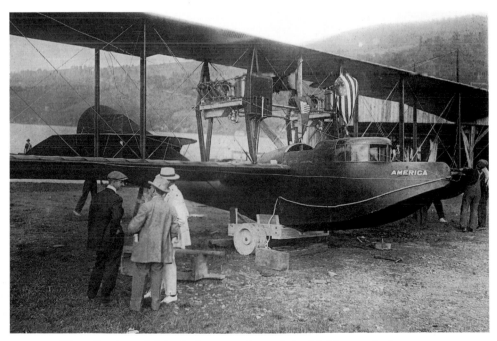

Glenn Curtiss, at left in the foreground, stands beside his 1914 airplane, *America*, designed to be the first aircraft to make a transatlantic crossing.

Thomas Baldwin (in the bowler hat near the center) stands below a Curtiss-powered dirigible (most likely the *California Arrow*) at one of the many exhibitions he gave before paying customers, circa 1907.

Orville *(left)* and Wilbur Wright, characteristically attired, in a 1910 photograph.

Curtiss *(right)* demonstrates to Thomas Baldwin the "wind wagon" he developed in 1907 to test new propeller designs for Baldwin's dirigibles.

A 1907 photograph of the Aerial Experiment Association (AEA) taken at Alexander Graham Bell's Beinn Bhreagh estate on Cape Breton Island in Nova Scotia. From left to right: Glenn Curtiss, Frederick W. "Casey" Baldwin, Dr. Bell, Lt. Thomas E. Selfridge, and J. A. D. "Douglas" McCurdy.

Lt. Thomas Selfridge sits at the controls of Bell's tetrahedral kitelike glider, *Cygnet II*, in a 1907 photograph taken near the Beinn Breagh estate.

The Aerial Experiment Association pictured here after their move to Hammondsport, New York, early in 1908. From left to right: Casey Baldwin, Lt. Thomas Selfridge, Glenn Curtiss, Alexander Graham Bell, and Douglas McCurdy. At the far right is Aero Club member Augustus Post, who was a frequent visitor to the group in Hammondsport and who would widely chronicle their work.

The AEA's third aircraft design, *White Wing*, near Stony Brook Farm in Hammondsport, New York. It was their first design to incorporate a wheeled tricycle undercarriage.

Glenn Curtiss's epochal *June Bug* flight, July 4, 1908, in Hammondsport, New York. As the first publicly announced flight in America, it won for Curtiss and the AEA the coveted *Scientific American* trophy for the first airplane to fly for a kilometer before judges.

Glenn Curtiss's 1909 aircraft *Rheims Rider* sits before the lavish grandstand at the *Grande Semaine d'Aviation* in Rheims, France. The exhibition, the world's first international air meet, would draw hundreds of thousands of spectators, including royalty from across Europe and many heads of state.

Curtiss's 1909 *Rheims Rider* rounds the pylon on a test flight for the James Gordon Bennett Prize. The top honor at the *Grande Semaine d'Aviation* in Rheims, France, the coveted Bennett Prize, is awarded to the world's fastest airplane.

Curtiss, with the cap near the center, sits with a young and eager crop of coworkers and protégés in Hammondsport, circa 1911. At the time, with the Wright brothers' legal battle against him in full swing, Curtiss relies heavily on income from exhibitions of his airplanes, and many of these younger pilots display his aircraft around the country.

This photograph, depicting the second phase of the reconstruction of the Langley aerodrome, shows the tandem-wing aircraft flying handily over Lake Keuka in Hammondsport, New York. A number of changes have been made to the aerodrome by this time, including the installation of a Curtiss engine, the addition of pontoons, and the raising of the pilot's seat above the aircraft's frame. At the helm for this particular flight in the fall of 1914 is Curtiss coworker Elwood "Gink" Doherty.

Almost immediately after Baldwin is launched, it is clear to all that the strong gusts are too much for the airship. The crowd watches dumbfounded as Baldwin and his craft begin to drift swiftly and inexorably away from the fairgrounds.

Curtiss and other assistants on hand immediately set out on a run. The Wrights, who by this time have had a good deal of experience with airborne mishaps, gamely skirt the cord roping off the spectators and join Curtiss and the others to try to recover the vast runaway airship, its tethering lines now dragging and flailing below.

What an improbable scene, like one of those parlor games where figures from history are imagined together in farfetched situations. As complete strangers, Curtiss and the Wrights run beside one another across the open fields of the Dayton fairground in common pursuit of a wayward aircraft.

The restraining ropes are quickly recovered, but the wind is so strong it takes the hard work of Curtiss, Wilbur Wright, and several others to tame the unruly dirigible. And as they tug the cumbersome airship back to its prominent spot at the fairground, a grateful Baldwin, disembarking none the worse for the ride, thanks them all for their indispensable and speedy assistance. After the strange ordeal, the Wright brothers formally introduce themselves to Curtiss as he catches his breath. And Curtiss finally—and eagerly— makes their acquaintance.

It makes sense that Thomas Baldwin would first bring together Curtiss and the Wrights. We think of the dawn of aviation as belonging to the enshrined airplane pioneers but, in the early years of the twentieth century, it was Baldwin and a handful of others who owned the skies. The vexing work of getting airborne, after all, was mostly shunned by the establishment. Travel in the air—such as it was with a colorful array of balloons, dirigibles, and boldly envisioned (and mostly impractical) heavier-than-air flying machines— was left largely to an extraordinary collection of outsiders, including cranks, charlatans, wealthy eccentrics, and showmen like Baldwin.

A balloonist and former tightrope walker in a traveling circus, "Captain" Baldwin—the title was a show-business honorific he gave himself—was already world-renowned when he burst into Curtiss's life in 1904. The avuncular Baldwin, brimming with worldly experience, would become an important influence on Curtiss, and would draw him into the world of aviation.

Like many others interested in flight at the turn of the century, Baldwin was galvanized by Santos-Dumont's flight around the Eiffel Tower. Aside from the impressive nature of the feat, Santos-Dumont, heir to a large, Brazilian coffee fortune, led an enticing life of glamour at the height of the Belle Époque in Paris, setting new trends in fashion and dining nightly at a regular table at the swank restaurant Maxim's. When Santos-Dumont's balloons and airships would get caught in the trees, friends such as the Rothschilds, would send up champagne lunches for him to enjoy during repairs. Louis Cartier even created the world's first wristwatch for him in 1901 so that he could tell time while keeping both hands on the controls of his dirigible.

For his part, Baldwin became determined to move beyond hot-air balloons to build the world's finest dirigible. His idea was to top Santos-Dumont's feat with even more dramatic flights in the United States—ideally before huge crowds of paying spectators. Baldwin financed the scheme with funds from two wealthy backers in California. Based on his balloon work, he had already perfected a formula to make the varnished silk for the gasbag. All he lacked was an engine light and reliable enough to power the machine.

As Baldwin would tell the story later, Curtiss's motor suddenly appeared before him like an auspicious omen. One day, in 1904, while Baldwin was building his dirigible on a ranch in California, a young man rode by on a new "Hercules" motorcycle—Curtiss's brand name at the time. Baldwin was instantly struck by the idea that the compact motorcycle engine was exactly what he needed to propel his new airship. He chased down the rider to learn where it had come from and immediately wired an order for an identical, two-cylinder motorcycle engine to the G. H. Curtiss Manufacturing Company in Hammondsport, New York.

In 1904, Curtiss's business was booming. He had only a handful of employees operating out of a cramped wooden factory building, but bicycles were selling briskly and the expansion into a line of motorcycles was nothing short of a sensation. Motorcycle orders were coming in so fast from around the country that customers faced a three-month backlog even with a stepped-up production schedule that included fifteen-hour days at the shop for many workers.

Around this time, Curtiss's wife Lena joined him at work, attending to the burgeoning paperwork, doing the bookkeeping, and helping in any way she could. The shop didn't have a typewriter but once or twice a week, Monroe Wheeler, a prominent local lawyer, let Curtiss or Lena come by to use his. A year earlier, the couple had lost their infant son to congenital heart failure. With Curtiss so busy at the shop, and with both of them deeply saddened by the loss, Lena's help at work made sense all around. The two seemed happy for the extra time together and it helped having such a competent extra hand at the shop. It was a somewhat unusual arrangement in Hammondsport at the time, but Glenn and Lena

Curtiss were both too industrious and practical to ever worry much about social conventions. Lena, the daughter of a local lumberman, was no stranger to work outside the home. She had been a grape picker—the seasonal work common for many teenage girls in the region—when she first met Curtiss.

By most accounts, Curtiss was always so distracted by things mechanical that he never showed any interest in girls before meeting Lena. But he noticed something about her. During their first encounter, when Lena was seventeen and Curtiss was just a year older, she offered him water while he was taking a breather from bicycling along a hilly country road on the outskirts of town. With brown, wavy hair and big eyes, she seemed as forthright, unpretentious, and basically shy as he was. Oddest of all to Curtiss as he came to know her better, she seemed to genuinely like and admire his interest in mechanical things. They were married within the year and remained unusually close throughout their lives together.

Five years into their marriage, when Lena began to work at her husband's manufacturing shop, she understood well that Curtiss's work was much more than just a job for him. He always seemed to be turning over technical problems in his mind and never seemed happier than when he was building something new. Working beside her husband suited Lena just fine.

According to one account, before they were married, Lena's father had been impressed by Curtiss. "That boy is going places," he liked to say, predicting that someday Curtiss "would have Lena living in a brick house." Clearly, Glenn Curtiss was going places. But she'd laugh when she would tell the story years later, adding that, for all the adventure she "never did get that brick house."

With a deluge of orders for motorcycles from around the country, it is not surprising that Glenn and Lena Curtiss did not pay much attention to an odd letter from a "Captain" Baldwin placing an urgent order for a motorcycle engine to be used in an airship. Baldwin's reputation as a showman may well have filtered all the way to Hammondsport, but it was not enough to keep Curtiss from being highly skeptical about the prospect of using one of his engines in flight. He did fill the order, however. With no two-cylinder engines on hand, Curtiss actually removed one from a recently minted motorcycle to ship to Baldwin. Curtiss was heard then and for sometime afterward referring to anyone wanting to take his engines aloft as an "aviation crank." But above all, he was a practical businessman: people could use his engines however they wished.

If Curtiss had doubts about flying, his views on aviation would change dramatically later in 1904 when Captain Baldwin showed up at his door.

Baldwin came by train to Hammondsport in the fall of that year directly after a stunning success with his dirigible at the Louisiana Purchase Exposition, a world's fair held in St. Louis, Missouri. The fair was an enormous event, designed to commemorate the centennial of the purchase of the Louisiana Territory. President Teddy Roosevelt even attended to review the opening parade. With a $100,000 grand prize for the best demonstration of an aircraft, the fair's organizers had drawn the attention of every aviator in the world, including Santos-Dumont.

Even the Wright brothers were tempted to display their flying machine. With the urging of their mentor Octave Chanute, the Wrights went so far as to visit the St. Louis fairgrounds and to ask the rules committee to make the contest guidelines more favorable to the inclusion of heavier-than-air machines. But ultimately, even the lavish prize money could not lure the Wrights into a public display.

With the Wrights' refusal to enter and with Santos-Dumont's dirigible unexpectedly damaged, it was Baldwin's airship—powered by a Curtiss engine—that won top honors after making a controlled flight over St. Louis. Mismanagement led the organizers to rescind their lavish reward, but even this didn't dim the acclaim Baldwin won. Banner headlines proclaimed him the leading airship designer of the era. And Baldwin attributed his success in great part to the craft's engine. Determined to meet personally with the manufacturer who had brought him such a triumph, Baldwin arrived in Hammondsport before Curtiss had even heard the news of his flight.

As Baldwin often explained afterward, he expected to find a big, important-looking industrialist at the helm of the Hammondsport manufacturing operation. He had a hard time believing that the lanky, unassuming twenty-five-year-old Curtiss, who sheepishly presented himself, was really the head of the firm. But if Baldwin was surprised by Curtiss, it is safe to say that Curtiss's amazement with Baldwin was greater.

It is little wonder that Curtiss would be enthralled with his world-renowned visitor. A hulking two hundred pounds, with a broad face and winning smile, Baldwin was a celebrity, certainly not the kind of visitor who graced remote Hammondsport every day. He seemed to have been everywhere, while Curtiss had hardly ever left Hammondsport. And best of all, under the circumstances of Baldwin's recent success in St. Louis, he didn't even sound too far-fetched when he predicted grandly that Curtiss engines would soon power fleets of airships around the world.

Curtiss immediately invited Baldwin to stay as his houseguest.

Lena couldn't help but be excited. She fretted about the simplicity of their home and made Curtiss wear his good suit to dinner. They were both grateful and relieved at how unpretentious Baldwin turned out to be. But it was his stories that charmed them the most: his seemingly endless stream of tales kept them rapt and often made them quite literally gasp in amazement.

As Baldwin recounted to the young couple, he had first taken his act to Europe in the late 1880s, after encouragement from world-famous, Wild West showman "Buffalo Bill" Cody. Baldwin's act was a roaring success and he performed his aerial feats before a long series of wildly cheering, sold-out crowds. On one of his European visits, he was the first foreigner Count Zeppelin ever took through the interior of the aircraft that bore his name. In London, the British Parliament had adjourned to see him perform. Baldwin even wore a diamond ring presented to him by Queen Victoria "for his daring, skill and aid to science."

Baldwin's personal history was as twisted and exciting as that of earliest aviation itself. Orphaned at the age of twelve, young Thomas Baldwin soon ran away to find his fortune. He sold newspapers in Missouri, worked as a street lamp gaslighter, tried his hand as a railroad brakeman in Illinois and followed that with a stint as a door-to-door salesman throughout the Midwest. After a chance encounter with a trapeze artist, Baldwin joined a traveling circus. He soon mastered the tightrope and received top billing for his fearless performances. In one of his most spectacular feats, Baldwin walked a high wire strung along the coast of San Francisco from the well-known Cliff House to Seal Rocks, some 700 feet long and 90 feet above the ocean.

In 1887, at the age of twenty-nine, Baldwin teamed up with a balloonist named Park A. Van Tassel and made a parachute jump from

Van Tassel's balloon before some thirty thousand paying spectators in San Francisco's Golden Gate Park. From then on, Baldwin was irrevocably hooked on aviation. Before long he began experimenting with his own hot-air balloon and his own design for a parachute with many visionary new features. It was made of pure silk, completely collapsible with flexible shroud lines and an eighteen-inch hole cut in the top for the air to escape through. The design was so effective that, even though Baldwin never patented it, he would long be credited with inventing the collapsible parachute in 1885.*

Baldwin's success in St. Louis and his visit to Hammondsport did much to change Curtiss's cautious skepticism about aviation. With Baldwin's encouragement, Curtiss plunged into designing engines specifically for use in dirigibles. Curtiss even constructed a land vehicle he called a "wind-wagon," replete with three wheels and propeller mounted in the rear, to test the power of the new motors and to help design an efficient aerial propeller. He took great glee in testing the machine on the outskirts of town. Much to the horror of his Hammondsport neighbors, the deafening prototype frightened farm animals and raised a cloud of dust in its sputtering wake. They wondered if Curtiss was getting carried away, quite literally, on a flight of fancy.

Curtiss even got his first taste of air travel by making an ascent in one of Baldwin's dirigibles in a pasture outside of Hammondsport. Theirs would be a lasting friendship. Ultimately, Baldwin would even move his operations to Hammondsport. He would make a total of thirteen airships outfitted with Curtiss motors in the years to come,

---

*Rigid parachutes, or aerodynamic braking devices, as they were sometimes called, developed considerably earlier, however. Their use dates at least to 1783 when, in the first well-recorded instance, Sebastian Lenormand used an umbrellalike contraption to descend safely after jumping from an observation tower in Montpelier, France.

including the first aircraft ever purchased by the U.S. military (aside from its abortive attempt to underwrite Langley's efforts).

Soon virtually all dirigible balloons operating in the United States would be driven by Curtiss motors. First prize at the St. Louis exposition in 1904 was followed by a dramatic and awe-inspiring flight over Portland, Oregon, at the Lewis and Clark Exposition in the spring of 1905. As was increasingly his penchant, the hefty Baldwin remained on the ground, sending in his place a fearless—and much more lithe—teenager named Lincoln Beachey, a man who would eventually become the most famous flier of his generation.

The Portland flight was so successful that Beachey decided to set out on his own. On a beautiful June morning in 1906, with no notice or official authorization, nineteen-year-old Lincoln Beachey took off from a park in Washington, D.C., in a dirigible outfitted with a Curtiss engine. He would be the first aeronaut to grace the nation's capital. He soared above the treetops, circled the Washington Monument, and then landed on the lawn of the White House, climbing out of the airship to meet an astonished first lady—Mrs. Edith Roosevelt—who interrupted a meeting to examine Beachey's contraption for herself. Awestruck, Mrs. Roosevelt rightly pronounced Beachey's stunt the most novel call ever made upon the White House and said she was sorry her husband Teddy was off giving a speech and not there to see it himself. With Mrs. Roosevelt's blessing, Beachey got back in the ship and breezed his way over to the Capitol Building, circling the dome several times as lawmakers flooded out from a joint session of Congress onto the Capitol steps for the spectacle. Beachey handily landed the dirigible before the spellbound leaders and sauntered over to spend the next hour answering their numerous and animated questions.

There was no doubt that the times were changing. Curtiss would still occasionally call his flying customers cranks, but like the rest of the world, he couldn't help but take the idea of air travel more seriously. Interest in flying moved further into the mainstream with the founding of the Aero Club of America in 1905. Scores of inventors around the world were actively trying to solve the puzzle of heavier-than-air flight. Their efforts would bring dramatic changes. And Curtiss would soon find himself in the center of it all.

How did the airplane come into being? Some inventions arrive at once, presenting themselves in a brilliant flash of insight. Curtiss's invention of the motorcycle's handlebar throttle is a small example. Still other developments occur by accident, often in the active pursuit of something else. Alexander Graham Bell, for instance, was working to improve the telegraph when he first stumbled on the concept of the variable resistance of electric current that would make the telephone possible.

The airplane, however, was an invention categorically unlike these fortuitous or accidental developments. On the contrary, it had been envisioned in rich detail for generations and was doubtless a mainstay of the imagination ever since humans first observed birds in flight. Leonardo da Vinci famously drew diagrams of flying machines as early as 1483 and some intellectual historians champion far earlier antecedents. The problem was making an airplane that would actually work. And the quest was a long one indeed.

Dreams and visions aside, the heavier-than-air flying machine was actively under development for a full century before the Wright brothers' success at Kitty Hawk. In fits and starts, with contributions from many sources around the world—some scientists and engineers, some daredevil nonconformists—the airplane's development moved incrementally forward.

Of course, these early efforts met with only limited success. Many pioneers displayed an incomplete understanding of aerodynamics. Others lacked a practical system of propulsion. Still other attempts at a working airplane led entire schools of researchers down blind alleys. Nonetheless, the standard engineering assessment that brands all these early efforts flatly as failures is a woeful misreading of history. Repeated and reinforced in scores of aviation texts and children's history books alike, such a view does more than miss a crucial truth about the origin of the airplane. With it we willfully deceive ourselves about the way new technology evolves.

Like most other technologies—and in fact like so many of humanity's great conceptual breakthroughs—the concept of the airplane percolated slowly, refined and distilled as many great minds grappled with different aspects of its deepening mysteries. In many ways, the airplane was like a difficult jigsaw puzzle; for it to succeed, many pieces had to be put together correctly. The puzzle would not be complete until its final piece was not in place. But many important victories would be achieved as seminal pieces came together along the way.

One extraordinary theoretical leap was achieved by Sir George Cayley, a British nobleman, who started his aeronautical investigations in 1796. Cayley was a gentleman scientist with a voracious and broad-ranging interest in the world around him. In his voluminous journal, he recorded all his measurements about the world around him, from the fact that crows flew 23 miles per hour on a calm day, to the observation that his thumbnail grew at a rate of exactly one-half inch in 100 days. He was the first to envision a practical airplane design as we know it today replete with fixed wings, a cruciform tail unit and a propulsion system. Perhaps even more important, Cayley deserves recognition as the father of modern aeronautics for his surprisingly sophisticated understanding of air resistance and the need to control an airborne craft along all its axes, now known as pitch, yaw, and roll.

A visionary who published his research and inspired many others to come, Cayley called the air "an uninterrupted navigable ocean that comes to the threshold of every man's door." At a time in the early nineteenth century when the view was nothing short of heretical, Cayley never doubted the viability of a heavier-than-air craft. As he put it in 1809, for instance, he was "perfectly confident" that someday people would transport themselves, their families, and their goods, "more securely by air than by water and with a velocity of from 20 to 100 miles per hour"—an audacious and startling prediction for a man of his day.

As important as his theories were, however, Cayley's work went far beyond the theoretical. In 1853, from the hill behind Brompton Hall, his British estate, Cayley launched the world's first full-scale glider capable of successfully carrying a passenger. Oddly enough, a half century before Kitty Hawk, Cayley's reluctant coachman was the first person ever to fly in a heavier-than-air contraption, soaring for some nine hundred feet over the British countryside and becoming so terrified in the process that he tendered his resignation immediately upon landing.

In retrospect, it seems clear that all Cayley lacked to make a working airplane was a lightweight engine. He quickly realized that the steam engines of his day were too heavy to do the job. And, after some brief experimentation with several alternative propulsion systems, he reluctantly gave up hope of motorizing his craft, once again writing presciently that a much cheaper "gaslight" engine would likely be produced someday. Given Cayley's precedent-setting place in history, decades before the internal combustion engine developed enough to be usable, that piece of the aeronautical puzzle would simply remain beyond his grasp.

Like Cayley, dozens of early aviation experimentalists played an indispensable role in amassing technical data about flight. Several of these, including Percy Pilcher in Britain, John Montgomery in California, and Otto Lilienthal in Germany, even lost their lives testing their designs.

Lilienthal's experimental work was undoubtedly the most important. A professional engineer who manufactured steam boilers, Lilienthal was passionate about flying and undertook a systematic study of the lifting power of surfaces and the movement of the center of pressure when wings are placed at different angles—an important step in understanding the stability of aircraft.

Often working after dark to avoid the opprobrium of his neighbors, Lilienthal recorded the results as he jumped repeatedly into the wind in the Rhinower Hills, six miles northwest of Berlin in hang gliders of his own design, soaring at times for more than one thousand feet. Based on his experiments, Lilienthal published detailed tables in 1889 calculating the lift of wings with different designs and camber, or pitch. Convinced that powered flight would one day be practical, Lilienthal successfully completed more than two thousand flights with his various gliders before he was killed in August 1896 when a sudden gust caused his glider to stall and crash to the ground.

Researchers like Lilienthal did not simply complete pieces of the airplane puzzle for themselves; they passed the information along to others. Before Lilienthal's death, for instance, Samuel Pierpont Langley traveled to Germany to see Lilienthal's gliders firsthand.

And the Wright brothers used Lilienthal's data extensively, especially in their earliest aeronautical research.

Some scientists were equally influential even though their expertise stood considerably afield from the engineering problem of how to build an airplane. One good example is the French naturalist Louis-Pierre Mouillard whose seminal 1881 work *L'Empire de l'Aire* carefully analyzes the wing structure and weight of bird species to systematically study how they fly. His seminal work influenced a generation of aviators. In fact, Langley attended an 1886 lecture by Mouillard and later credited it as the impetus for his decision to pursue aeronautical work in earnest. Similarly, the pioneering French inventor Clement Ader—one of the world's very first to build a heavier-than-air flying machine—was so taken with Mouillard's approach that he trekked through the wilds of Algeria in search of large vultures, luring the huge birds with slabs of meat to personally inspect the way they soared.

But for all the rich exchange of aeronautical information, no one during this period would match the efforts of Octave Chanute. His 1894 *Progress in Flying Machines*—still in print today—may well be the most important work in the history of aviation. It presents a panoramic picture of the emerging aviation field near the turn of the century, and was read by virtually all of the earliest aviation pioneers around the world at the turn of the century—including the Wright brothers.

Chanute's book offered detailed and precise descriptions of the work of an extraordinary collection of no fewer than sixty-five inventors with glider or airplane designs as of 1894—from the pathbreaking box-kite designs of Lawrence Hargrave in Sydney, Australia, to the rounded, birdlike gliders of the French ship's captain Jean-Marie le Bris.

To anyone seeking to understand the origins and development of the airplane, then or now, the existence of Chanute's community of researchers at the turn of the century is a source of wonder and fascinating technical information. These aviation pioneers whose work Chanute chronicles were often marginalized and subjected to ridicule. But they represented an emerging understanding of aeronautical engineering that was essential to the airplane's development.

It is astonishing to remember that, even before the Wrights' success at Kitty Hawk in December 1903, some seven stalwart experimenters (including Langley) each managed to build motorized, full-scale heavier-than-air flying machines, and get them into the air before witnesses for at least a hop. The list includes little-known names like the French engineer Félix du Temple de la Croix, Russian inventor Alexandr Fyodorovich Mozhaisky, and the Austrian piano maker Wilhelm Kress.

One member of this group, the expatriate American Hiram Maxim, famed for his invention of the machine gun, spent roughly $100,000 of his own fortune on a massive, four-ton flying machine at his estate in Kent, England. On July 12, 1894, Maxim briefly flew the gargantuan aircraft, powered by two hulking 180-horsepower steam engines that each turned a propeller nearly eighteen feet long. Maxim set the machine upon a steel rail and, in order to measure the machine's lift, built a guardrail to prevent it from getting more than a few inches off the ground. But the machine had so much thrust that, with Maxim at the helm, it immediately rose high enough to break through the iron guardrail, twisted out of control, and crashed. Maxim was so unnerved by the experience that he soon abandoned the entire project.

Maxim never made a fully working aircraft. But his assessment in

1890 gives a clear sense of the way things stood in the field. As he put it, "I think I can assert that within a very few years someone—if not myself, somebody else—will have made a machine which can be guided through the air, will travel with considerable velocity, and will be sufficiently under control."

After Baldwin's wayward dirigible flight at the Dayton Fair, he invites the Wrights aboard the catwalk as thanks for their assistance. They eagerly accept the offer and reciprocate by inviting Baldwin and Curtiss back to their Dayton shop. The four aviation entrepreneurs strike up a natural conversation about the emerging field and talk late into the evening. They talk generally about the state of technological development in aviation. They talk about propeller design, a topic of considerable recent interest to Curtiss.

Curtiss is very much at home in the Wrights' shop. With workbenches, tools, and bicycle rims hanging from the rafters, it is remarkably like his own in Hammondsport, only perhaps a bit more fastidiously kept. Curtiss himself bears such a resemblance to the Wrights he could almost pass as another Wright brother himself. And they share many features in background and temperament as well. All three ran bicycle shops. None of them continued their formal education past the eighth grade. All tend to be shy and reserved by nature, especially with strangers. Most strikingly, all three are expert at a hands-on kind of thinking—making adjustments and tinkering to improve their products. It is an indispensable trait in the young aeronautical field where so much still needs to be discovered by feel.

In hindsight, of course, it is equally notable how well their special talents complement one another. Curtiss excels at engine design whereas Orville will later acknowledge that the engine was always the weakest component of the original *Wright Flyers*. But despite Curtiss's formal offer to collaborate with the Wrights by building them a special engine, history will dictate otherwise.

Curtiss's offer to collaborate—and the Wrights' prompt refusal—also highlights a dramatic difference between Curtiss and the Wright brothers. Despite his shyness, Curtiss is also open and candid; he loves nothing more than to bounce ideas back and forth with friends and colleagues. His inclination is always to work in a group and he thrives when exchanging his ideas with others.

By contrast, Wilbur and Orville Wright operate as a self-contained unit. Sons of a bishop in the United Brethren Church, they have been trained from an early age to be scrupulously truthful but, throughout their lives, they tend to be guarded and mistrustful of outsiders. Even though theirs is now a three-year-old invention, legally protected by a U.S. patent, the Wrights never consider taking their interested visitors to the nearby shed where their airplane sits hidden. It is a testament to how positively they feel toward Curtiss and Baldwin that they let the two see a picture of their airplane in flight that evening.

Sadly, much later, when relations between Curtiss and the Wrights become strained, Orville Wright will offer the baseless charge that the chance meeting in Dayton began a systematic effort by Curtiss to steal the Wrights' technology. The visit couldn't have helped but to encourage Curtiss about the potential of heavier-than-air flight, but the charges otherwise couldn't be farther from the truth.

Curtiss's open eagerness to engage the Wrights shows in his correspondence immediately following the visit and for some time afterward as well. When he returns to Hammondsport, Curtiss sends a warm and chatty letter to the Wrights offering details of his work on a new eight-cylinder motor and other exploits. "It may interest you to know," he writes, clearly following up on an exchange in Dayton, "that we cut out some of the inner surface of the blades on the big propeller, so as to reduce the resistance and allow it to speed up, and it showed a remarkable improvement." It is hardly the correspondence of someone trying to surreptitiously appropriate the Wrights' ideas.

Baldwin's last day in Dayton is the most successful. To the delight of many spectators, he makes a dramatic flight above the city, traveling from one end of town to the other and back, and making a perfect landing in time to beat an automobile that tries to race the same distance along the city streets.

Baldwin thus adds his feat in Dayton to a remarkable string of airborne accomplishments. Now mostly forgotten, Baldwin dominated the air in the brief era of the dirigible. And he continued to entertain crowds with his aerial feats as the airplane blossomed, becoming the first person to hold a complete set of international licenses to pilot all three types of aircraft: balloon, dirigible, and airplane. But Baldwin's moment in the limelight of aviation soon faded as his showman's sensibility was eclipsed. Aerial exhibitions continued for many years to come, but a new generation of aviators like Curtiss had even more ambitious goals for the airplane.

FIVE

SKY DANCING

The age of the flying machine is not in the future.
It is with us now.

—Alexander Graham Bell, 1906

With one of his motorcycle motors safely packed in the baggage car, Glenn Curtiss gazes out the window of a northbound train at the open blue sky. It is a cloudless morning in early July 1907. Curtiss will spend most of the next two days on trains, including many stops and changes, but he is happy for the solitude. It is his first stretch alone for a while, and the remarkable nature of his voyage has put him in an uncharacteristically pensive mood. He is quite sure he is either losing his common sense completely or embarking on one of the most exciting adventures of his already eventful life. He just cannot decide which.

Curtiss is en route to Alexander Graham Bell's summer estate on Cape Breton Island, Nova Scotia. It will be one of the farthest voyages from Hammondsport that Curtiss, at age thirty, has ever taken. He has left his successful, growing motorcycle and engine manufacturing business in the capable hands of Henry Kleckler and Harry Genung. The shop now boasts more than two dozen employees and its operations have steadied out considerably by this time; the orders keep coming at a startling rate, but now they are more speedily and reliably filled. Nonetheless, Curtiss wonders, is this a time to get involved in something new, risky, and untried?

As the train turns eastward from Montreal toward the Atlantic Ocean and Canada's easternmost province, Curtiss lets his mind wander. He thinks about his entry into the motorcycle business. Years earlier, Curtiss had first sent away for a set of castings with which to make a gasoline engine. They had arrived with no instructions. Through trial and error, Curtiss had machined the pieces and assembled his first makeshift engine. Tomato cans had served as both carburetor and gas tank.

An almost imperceptible smile crosses Curtiss's face as he remembers his first trip to the post office after rigging the engine to a bicycle, using a leather strap to drive the wheels. He had to pedal all the way across town while Hammondsport shopkeepers stood in their doorways and laughed at his awkward contraption. But, as he liked to recount later, the engine finally began to spark and pop; then it was his turn to laugh. Before long, everybody in Hammondsport wanted a motorcycle from Curtiss, and orders began pouring in from all over the country.

If Curtiss prides himself on anything, it is his readiness to adopt and work with new technology. In 1901, in the days before electricity had come to town, Curtiss had installed acetylene gaslights in many of the shops around Hammondsport's town square. And he had designed a clever system with tin cans and a few dabs of solder to make them burn more efficiently.

For as long as he could remember, Curtiss had never sat on his hands in the face of opportunity. As early as age fourteen, he displayed this trait, in one of his first jobs at the Eastman Kodak works in Rochester. Along with many other boys his age, Curtiss stenciled the numbers on the film that would show through the red window in the camera's back. Soon after taking the $4-per-week job, Curtiss went to his boss with a plan. He asked the company to pay him and his coworkers by the piece, at a rate of 25 cents per 100 strips. With the workers averaging 250 strips a day, the piece rate he proposed was roughly equivalent to their $4 wage.

As soon as the company accepted his proposal, Curtiss brought in a rack he had designed that could hold a pile of a hundred strips, a hinged mechanism to hold the stencil, and a fat new brush with which to dab on the ink in one stroke. He then showed his mostly teenage coworkers the way to riches. Before long, Curtiss had upped production tenfold. He and the other boys were going so fast, they earned more than some Kodak managers. Even when the company renegotiated the deal down to 9 cents per 100 strips, Curtiss's ingenuity still meant that the team had almost tripled their weekly wage.

Now the eminent inventor Alexander Graham Bell has asked Curtiss to join forces to engineer a heavier-than-air flying machine. But Curtiss wonders whether he can bring his ingenuity to bear in any meaningful way in a complex and unknown field like aeronautics. His engines helped Thomas Baldwin and his dirigibles in ways Curtiss never anticipated. Bell might open similar new horizons. But the prospect of making heavier-than-air flying machines a practical reality seems far-fetched, despite the Wright brothers' claims and Bell's enthusiasm.

There is also the question of Bell himself. He and Curtiss first met a year and a half before, in January 1906, at the New York City

Auto Show. It was hard not to be impressed with Bell, a charismatic and warm man with bushy, white whiskers, a booming baritone voice, and a gentle, learned demeanor.

For years, Curtiss had found the auto show a good forum for drumming up motorcycle business. That year, though, the show had notably changed when organizers invited the city's fledgling Aero Club to participate. Curtiss had prepared his customary exhibit, adding emphasis on dirigibles in honor of the new participants. But, if anything, he had underestimated the crowd's new-found interest in aviation. The exhibits committee had appealed to Langley, Chanute, and Bell for examples of their aeronautical work, and all had lent gliders or models to the exhibition. No fewer than four dirigibles, including Baldwin's latest, hung suspended from the ceiling of the large exhibition hall. The committee had also appealed to the Wrights, but, not surprisingly, the brothers sent only the tiniest offering: the exceedingly plain crankshaft and fly-wheel of their 1903 engine. "It would interfere with our plans," they wrote the committee, "if we should make public at once a description of our machine and methods."

At the 1906 show, no one was more in demand than Bell. He mounted a large exhibit of the striking, multi-celled, tetrahedral kites with which he was experimenting as a way to carry humans into the air. Bell also gave the keynote address at the Auto Club's banquet, declaring that the age of what he called "the aerodrome" was at hand. It wouldn't be long, he said, before aircraft would fly across the Atlantic. The day will come, he told the mostly skeptical crowd, "when we will leave New York City in the morning and be in London at night."

During the show, Bell visited Curtiss's exhibit. He was deferential and enthusiastic as he closely inspected the motors on display.

The machinery clearly captured the older man's interest, and he would make special note of them in his diary following the show. Not long after, he began to refer to Curtiss as "the greatest motor expert in the country."

Curtiss was unaware of it, but Bell was hatching a plan to gather together a team to try to build a workable flying machine. Sixty years old, Bell had learned a lot about the process of technological innovation, but, given his age, he also knew any team he established would have to include the best young and energetic talent he could find.

Bell hinted at his plan when Curtiss met him the following spring at Bell's stately home in the Georgetown section of Washington, D.C. In town on business, Curtiss had come calling and received a warm and enthusiastic reception. Bell treated him to a long and animated discourse on heavier-than-air flight that left Curtiss fascinated despite himself. Curtiss had long been wary of believers in flying machines, and even his experience with Baldwin hadn't fully shaken him of the prejudice. Yet Curtiss was enticed by the encyclopedic Bell and his mostly unfamiliar, firsthand stories of the aviation pioneers Maxim, Langley, and Chanute.

Curtiss had been exposed to the world of aeronautics through Baldwin, and he had even met the reclusive Wright brothers. But he had little idea of the extent of flight research that was being conducted around the world. It was hard not to be seduced by Bell's enthusiasm for the subject. To hear Bell tell it, the airplane's time had come, and it would soon burst upon the world the way the automobile and the telephone had only a few years before.

Not long after their visit in Washington, Bell came to Hammondsport to try to clinch a deal with Curtiss. As he purchased one of Curtiss's most powerful engines, he formally requested that Curtiss hand-deliver it to his summer laboratory in Nova Scotia. That way, he said, Curtiss could see his operation and help him and his assistants learn to operate and maintain the engine and adapt it to their needs. As a further enticement, Bell offered to pay Curtiss $25 per day, plus expenses, for his services, a handsome consulting fee at the time.

During that brief, introductory visit to Hammondsport, Bell thoroughly charmed Glenn and Lena Curtiss. Bell was a notably unpretentious person, especially considering his extraordinary accomplishments. That quality particularly appealed to Lena. She was both proud and a little taken aback that such a famous inventor would think so highly of her husband's motors that he would come to buy one personally for his aeronautical experiments. After all, this was a man who had utterly transformed the world once already with his invention of the telephone. Might he do it again, she wondered, with a Curtiss motor at the heart of his invention?

Bell's enthusiasm and generosity left little question that Curtiss would accept his invitation. He felt honored by Bell's interest and he would be well paid for his time. No harm could come, Curtiss decided, from taking a few days away from business to see for himself what Bell was up to. Like Lena, Curtiss enjoyed Bell's attention and appreciated his worldly experience. The effect of his visit was much like that of Baldwin's several years before, but even more heady. Where Baldwin was flamboyant, Bell seemed substantial. Ever the showman, Baldwin, when he prepared to leave town, liked to broadcast the fact that he was off to perform at some distant fair or exhibition. Bell was different. Upon his departure from Hammondsport, he only reluctantly divulged that he had to make a detour to England before returning to his beloved estate in Nova Scotia. When Curtiss asked about the British trip, Bell explained, almost sheepishly, that Oxford University was awarding him an honorary degree.

After endless stops and changes of train, Curtiss arrives at a little wooden station in a town that makes Hammondsport seem like a metropolis. Two ruddy, casually dressed young men greet him heartily, full of smiles. They are F. W. "Casey" Baldwin—no relation to Captain Baldwin but rather the son of former Canadian Premier Robert Baldwin—and Douglas McCurdy, the son of a neighbor and old friend of Bell's. As Curtiss soon learns, both young men are Bell's technical assistants, recent graduates of the University of Toronto with master's degrees in engineering. Much like their mentor, they are filled with enthusiasm, intellectual curiosity, and technical knowledge. They make a winning welcoming committee and, from the start, ply Curtiss with all manner of questions about himself and his motors.

The two eagerly help Curtiss haul his engine off the train and carry it a few blocks to a rickety pier. From here, a small steamer ferry takes them across the vast Bras d'Or Lake to Bell's one-thousand-acre estate called Beinn Bhreagh (from his native Gaelic for "beautiful mountain"). The boat sets off as soon as they board.

The Bras d'Or, a large, salty inland sea, covers much of the interior of Cape Breton Island. Even though he grew up in the beautiful Finger Lakes region, Curtiss can't help but marvel at the scale of the unspoiled scenery. After about a half hour, the ferry approaches a small pier that juts out from pine woods into the lake. Through the trees, Curtiss can see the windows and gables of a great rambling house rising high on the hillside above him like a big, shingled castle with turrets and parapets.

On the path leading up to the house, Curtiss passes some of the buildings that make up Bell's summer laboratory. Long, equipment-filled sheds stand close behind the boathouse. A long, narrow building is called the Kite House. Through its barn doors, Curtiss sees a baffling assortment of large kite-like structures hanging from the rafters, many covered in deep red silk.

Leaving the motor on the pier, the three make their way farther up the hill to the great screened-in front porch of the mansion, a lovely, breezy room with a majestic view across the water. Casey Baldwin and McCurdy introduce Curtiss to the household. All its members have gathered and now rise from comfortable wicker chairs to welcome him. Curtiss is greeted warmly first by Bell and his wife, Mabel. Also visiting are Lieutenant Thomas Selfridge, a U.S. Army aviation expert, invited by Bell to observe his aeronautical experiments; Bell's two daughters, Elsie May and Daisy; Daisy's husband, a botanist named David Fairchild; and their two lively grandchildren.

They graciously set Curtiss up in one of the finest guest quarters, a circular room in one of the corner towers. Baldwin and McCurdy escort the visitor and his small suitcase up to his room, and, as he sets it down next to the bed, Curtiss pauses to gaze wordlessly at the breathtaking view of the huge, sparkling lake below.

That evening, after an animated dinner in the Bells' cozy, wood-paneled dining room, the group gathers before the immense front hall fireplace for a lively session of talking and singing the likes of which Curtiss has never before encountered. Bell loves to sing and has an excellent voice; both he and Mabel play the piano. But even more than the jovial atmosphere, Curtiss is taken with the breadth and erudition of the conversation.

In this regard, no one amazes Curtiss more than the lady of the house, Mabel Gardiner Bell. Mabel is every bit as intellectually alive as her husband, and she fully holds her own on almost all the wide-ranging topics of the evening. She is also a suffragette: as early as 1901, she helped convince Alec, as she calls her husband, to champion the idea of universal suffrage—for women and blacks. And she is as adventurous as she is well educated. Some years earlier, she even went underwater in a diving bell as Alec watched in admiration from the surface, flatly refusing to try it himself.

Curtiss can see that Mabel clearly adores her husband and humors him in many ways. But she also teases him mercilessly, a trait that only adds to the evening's relaxed intimacy. At one point she humorously warns the others about their host: on their honeymoon, she recounts, her new husband took all the sugar cubes at the table and dropped them, one by one, into his coffee because he got curious about the tiny bubbles they produced.

Mabel is also deaf, having lost her hearing to scarlet fever when she was a small girl. In fact, it was Bell's desire to build an electrical device to help her that led to his invention of the telephone. As a good speaker and excellent lip reader, her handicap hardly slows her down.

Curtiss is particularly sensitive to her situation because his own sister Rutha had lost her hearing to meningitis as a young girl. As a result, Curtiss had learned both sign language and the habit of clearly enunciating his words so that his lips could be easily read. This knowledge proves useful before that first evening is over. At one point, Curtiss notices an unhappy expression creeping onto Mabel's face, and he realizes that she has lost the thread of the conversation. Quietly turning toward her, he deftly signs the missing words with his fingers. Her eyes light at once with surprise and gratitude, and the two form a special bond from that time forward.

So begins a week of exploration and camaraderie that is a revelation to Curtiss; Bell's home provides a creative atmosphere that falls somewhere between the rigors of a fast-track engineering laboratory and the playful pleasures of a child's summer camp. During the day, Curtiss works closely with the rest of the tight-knit group, running propeller tests and teaching them about the functioning and maintenance of his engine. The time passes swiftly as the men work in the lab, talking incessantly about airplane design and testing Bell's tetrahedral kites. Ultimately, Bell hopes to install Curtiss's engine in a large version of the kite in an effort to fly it on its own power with a pilot.

Around sundown, the group normally retires to the porch or the great hall to play with the children and drink hot tea. Evenings mean long and stimulating discussions punctuated, at the Bells' irresistible urging, by singing. The group talks about everything from atmospheric pressure and the inherent strength of tetrahedral construction to the mechanics of propeller torque. As often as not, Bell draws upon his vast intellectual passions, bringing into the mix everything from his early experiments with the telephone and telegraph to genealogical deductions about gender and disease drawn from his card-indexed classification of seven thousand members of a randomly chosen family called the Hydes.

As the evening hours wear on, some indefatigable subset of the group invariably adjourns to the study where Bell, smoking his pipe, opens his voluminous notebooks crammed with wide-ranging ideas and sketches. Bell always retires last; his preference is to work in the quiet of the night and sleep until noon. As a result, he also frequently chooses to muffle his telephone with towels, though it invariably manages to wake him nonetheless. "Little did I think, when I invented this thing, that it would someday rise up to mock me!" he exclaims to Curtiss one disgruntled morning on a separate occasion.

While Curtiss is the only team member at Beinn Bhreagh without an advanced degree, his mechanical genius and practical engineering experience offer a vital and formerly missing element. In this sense, Bell's plan begins to come to fruition. In anticipation of Curtiss's visit, Bell had noted in his journal that Curtiss's input would create a group with the "ideal combination for pursuing aerial researches." Now, during Curtiss's stay, his journal entries reiterate the sentiment more emphatically. His elation is clear: "I now have associated with me gentlemen who supplement by their technical knowledge my deficiencies; in this combination I now feel that we are strong, where before we were weak." Best of all, he writes, all members of the group are united by a common desire: "to get into the air."

On Friday, July 19, Bell invites the group into his study after a morning of experiments to formally discuss the possibility of forming an aviation enterprise. He has already broached the plan with each of them individually, but this is the first time they meet together to consider the matter in detail. With only a brief break for dinner, the lively meeting continues late into the evening.

For his part, Bell offers to conduct the experiments under his auspices and to raise the seed money necessary for the task. As Bell explains, Mabel has generously offered to sell a piece of property she owns in Washington, D.C., in order to supply the business with $20,000 in start-up capital. Further, Bell offers the use of his labs and guarantees to supply another $10,000 of capital, if required. In return, he proposes that he and Mabel will share a controlling interest in the new association.

Even beyond these logistical details, Bell has thought a good deal about the role he anticipates in the association. As he writes in his journal: "My special function, I think, is the coordination of the whole—the appreciation of the importance of the steps of progress and the encouragement of efforts in what seem to me to be advancing directions."

In mulling over Bell's proposal, Curtiss tries not to let the romance of Beinn Bhreagh cloud his business sense.

Is Bell close to inventing a flying machine? His tetrahedral kite-like gliders certainly seem strange, but it is hard to tell how they might fare with a motor attached. One question is the extent to which Bell's dramatic earlier success with the telephone—success that brought him fame and fortune by the time he was thirty years old—relied on the special circumstances of his background.

From the time of Bell's birth in 1847, his father had been well established in Edinburgh teaching speech and elocution and beginning research on what he called a "visible speech" system: a means to represent with symbols all possible positions of the vocal organs and the sounds to which they corresponded. George Bernard Shaw, a family friend, would immortalize Bell's father and his speech research as Professor Henry Higgins in the play *Pygmalion* (later adapted into the musical *My Fair Lady*).

From the earliest age, Bell had been further influenced in his interest in speech and acoustics by the fact that his mother was nearly deaf and, even as a young boy, Bell, of all his family members, could best communicate with her. After studying at the University of London, it was only natural for him to become involved in teaching speech to deaf pupils. For the rest of his life Bell considered this his core profession and always identified himself as a teacher of the deaf.

In 1871, when Bell was twenty-four years old, he moved with his parents from Edinburgh to Ontario, Canada, and then soon moved again, this time alone, to accept a professorship at the recently opened Boston University. There he introduced Helen Keller to her life-changing teacher, Anne Sullivan, a fact that would lead Keller to dedicate her autobiography to him. "Child as I was," Keller would write, "I at once felt the tenderness and empathy which endeared Dr. Bell to so many hearts. That interview would be the door through which I should pass from darkness to light."

To be sure, by the time Bell set out on his telephone research, he had developed a remarkable and unusual constellation of talents, including a mastery of aspects of speech, music, acoustics, mechanics, and electricity. The big question for Curtiss is whether Bell can bring his many talents to bear effectively on the field of aeronautics.

In this regard, Curtiss's time in Nova Scotia has shown him that there is much more to Bell's capacious repertoire than the telephone.

In his Georgetown lab, for instance, Bell developed the graphophone, a machine that used cylinders of hardened beeswax to replay voice recordings. Edison bought the rights and used Bell's work to help develop his phonograph. Among many other lesser-known accomplishments, Bell even invented a portable desalination device for use by shipwrecked sailors.

In one episode illustrating the breadth of his work, the White House called upon Bell when U.S. President James Garfield was shot in the back by a disgruntled federal worker in 1881. The bullet had lodged in the president's chest but did not kill him, and Garfield's advisers asked Bell to see if he could devise a means to help surgeons locate the bullet. Working frantically, Bell met the challenge by developing a crude metal detector using induction coils and a telephone. The machine worked brilliantly, but Bell was stymied in trying it on the ailing president when he couldn't make the machine stop buzzing. He never imagined that Garfield's state-of-the-art, high-tech mattress contained metal springs, and no one in the White House had known to mention it. Within days, Garfield died from the infected wound, despite Bell's frenzied and creative efforts.

Not long afterward, when Bell's son, Edward, died of a collapsed lung soon after birth, Bell combated his grief by developing a "vacuum jacket" apparatus to help patients with similar respiratory ailments to breathe. That invention predated the so-called iron lung by some four decades.

Now Bell is devoting his formidable inventive attentions to the challenges of flight. He will, Curtiss feels sure, bring a vast knowledge of many related scientific fields to the enterprise. He has traveled the world and remained in close contact with an extraordinary collection of colleagues from scientific and high-society circles. He is also an impressive polymath who seems to read everything, keeping voluminous notebooks and scrapbooks of clippings drawn from a mind-boggling array of American, British, French, and German publications.

Even more attractive to Curtiss than the breadth of Bell's knowledge and contacts is the simple fact that, after Bell's extraordinary initial success with the telephone, he has never stopped inventing and is wholly unafraid to apply his talents to wildly disparate ventures. He is an energetic and independent thinker who has made his own, distinct route through the world. Bell's many successes attest to the fact that he can bring vast experience in the day-to-day process of technological invention that will be indispensable to the group.

If he ever doubted it, Curtiss's extended visit helps fully persuade him of Bell's brilliance. Despite Bell's modest diary entry about being merely a coordinator of the project, Curtiss recognizes that he is a remarkable inventor, one of the true scientific and technological powerhouses of his day. Equally impressive is the undeniable way he has thrown himself wholeheartedly into the project. Even at age sixty, with a full life and a long record of achievements behind him as well as ongoing projects, Bell is embarking with gusto upon the task of designing a flying machine.

In addition to all these assets, Curtiss adds the fact that aviation is Bell's most enduring interest aside from speech. Even his telephone assistant Thomas A. Watson recalled, "From my earliest association with Bell he discussed with me the possibility of making a machine that would fly like a bird. He took every opportunity that presented itself to study birds, living or dead. . . . I fancy, if Bell had been in easy financial circumstances, he might have dropped his telegraph experiments and gone into flying machines at that time."

Bell's passion to build a flying machine was only heightened through his close friendship with Samuel Pierpont Langley. Thinking of Mabel's story of Bell's unswerving attention to the sugar cubes at his honeymoon breakfast table, Curtiss considers it an undeniable asset that, as Mabel now likes to remark, her husband Alec is "just gone about flying."

At the end of his weeklong visit, Curtiss is sorry to leave the remarkable, close-knit group. He returns to Lena in Hammondsport with a new outlook about the prospect of developing a working airplane—and also, perhaps, with a new sense of what it means to be an inventor.

Curtiss's style and his approach to engineering problems in the future will be forged by his association with Bell's team. Before meeting Bell, Curtiss had harbored a deep skepticism not only about aviation but about inventors in general.

"I used to resent being called an inventor," Curtiss said much later. "An inventor, as people in country towns thought of him, was a wild-eyed, impractical person, with ideas that wouldn't work. Perhaps I got some of that impression from J. T. Trowbridge's poem, 'Darius Green and his Flying Machine.' My grandmother knew Mr. Trowbridge very well, and used to recite that poem to me as far back as I can remember." The popular poem, published in booklet form with comic illustrations, lampooned an inventor's harebrained efforts to fly. Trowbridge's characterization, Curtiss said, simply reflected that, in terms of public esteem, inventors "didn't stand very high in rural communities."

Curtiss's visit to Beinn Bhreagh all but erases this childhood prejudice. His time at Bell's estate was one of the most exciting experiences of his life. But he still worries about taking time away from his rapidly expanding business to explore aeronautics. Ultimately, according to at least one source, Lena encourages her husband to continue with Bell, counseling him to trust his instincts and Bell's spectacular earlier success with the telephone.

Within a week of returning to Hammondsport and receiving Lena's blessing, Curtiss cables Bell to express his interest in joining the group. "I have given the Association plan considerable thought," he writes, "and am very favorable toward it."

When he learns that Captain Baldwin has an engagement to exhibit his dirigible in Halifax, Curtiss uses the event as an impetus to schedule a return visit to Beinn Bhreagh at the end of September. With logs blazing in the cavernous stone fireplace of the great hall, the group formally votes to establish their association.

After lengthy discussions, they amicably settle most of the particulars. They decide, for instance, that each person will get a chance to test his own aeronautical design. But their first effort will be to finish Bell's pet project, a piloted, tetrahedral kite.

Bell reiterates that, as an association dedicated to experimentation, he expects that there will be little or no profit, although any earnings will be divided equally among the partners. In her role as a funder, Mabel says, she will advance funds up to the amount of $20,000. As a show of gratitude, the group decides that, for each $1,000 she contributes, she will be assigned a 1 percent interest in any proceeds that might result from the group's work.

The group also settles on titles and salaries for the members: Bell is appointed the association's chairman with no salary. As a commissioned officer in the U.S. Army, Lieutenant Thomas Selfridge, appointed secretary, also declines a salary. Casey Baldwin will serve as chief engineer and Douglas McCurdy is named treasurer. Each of the young men will receive an annual salary of $1,000. Reflecting his status and esteem within the group, Curtiss is formally appointed director of experiments, with a salary of $5,000. He says he will draw his salary only when his business frees him enough to participate fully in the aeronautical work.

The next day, October 1, 1907, the group travels 230 miles to Halifax to make their new organization official. There, before a notary public and in the presence of the U.S. consul to Nova Scotia, the five members and their benefactor Mabel sign an agreement spelling out a yearlong arrangement. With this earnest and auspicious start, the Aerial Experiment Association, or AEA, is officially formed.

As soon as they are through, the group celebrates with a trip to the Halifax fairgrounds to see Captain Baldwin fly his dirigible. The cool fall day offers fine and clear weather for flying and, as usual, Captain Baldwin puts on an impressive show, piloting his dirigible in controlled circles above the avid spectators. Bell is delighted. For all his interest and enthusiasm in aviation, it is the first piloted, motorized flight he has ever witnessed.

With everyone in high spirits that evening, Captain Baldwin joins the group at a dinner hosted by the Bells at the Halifax Hotel to celebrate the new association. Bell toasts Curtiss and Captain Baldwin for the aeronautical accomplishments they have already achieved.

Never much of a speech maker, it testifies to his ease with the group that Curtiss now offers up a formal statement noting that he is "honored to have the opportunity to associate myself with Dr. Bell and the other members of the Association." With one big step, he has leaped into the vast unknown: the quest to design and build a working airplane.

Back in Hammondsport, in a swirl of excitement, Curtiss again contacts the Wrights. His inclination is to cooperate in some fashion, and he offers this in a letter to the brothers on December 30, 1907. His tone is deferential and warm. He tells the Wrights of his role with the newly formed AEA. He invites them to visit Hammondsport. And he offers to provide them his latest V-8, 40-horsepower engine for free.

Wilbur writes back declining both offers. But his tone is still friendly. "We remember your visit to Dayton with pleasure," he notes. "The experience we had together in helping Captain Baldwin back to the fairgrounds was one not soon to be forgotten."

Less than a month later, in correspondence with Octave Chanute, Wilbur writes that, despite the stepped-up efforts of aviation researchers like Curtiss, he and Orville are confident that "an independent solution to the flying problem is at least five years away." But Wilbur underestimates Curtiss and his new team. In an extraordinary collaborative effort, with Curtiss's superior engine, Bell's shrewd oversight, and many of the pieces of the aviation puzzle falling into place, the AEA will independently develop a working airplane in just five months.

SIX

FLIGHT OF THE *JUNE BUG*

What a moment for the vivid imagination.
The thing is done. Man flies!
                    —DAVID FAIRCHILD, JULY 4, 1908

Just past dawn on July 3, 1908, Glenn Curtiss, Lieutenant Thomas Selfridge, Douglas McCurdy, Casey Baldwin, and Henry Kleckler venture out to a makeshift hangar on the outskirts of Hammondsport. Their excitement mounts as first light streams across the verdant farmland around them.

Since midwinter, when the Aerial Experiment Association moved its operations from Nova Scotia to Hammondsport, many in town have realized that Curtiss and his team are on to something extraordinary. Despite the early hour, about a dozen of Curtiss's neighbors have trekked the two miles from Hammondsport to witness a spectacle. The team members stand quietly now on the edge of Stony Brook Farm. Here, beside a large potato patch, Harry Champlin, founder of the Pleasant Valley Wine Company, has built a half-mile racetrack for his horses. And, while the word has yet to come into common parlance, this morning the track will serve as a runway.

The team gingerly rolls the new fragile contraption they call an aerodrome out from its makeshift tent. To the assembled spectators, it is a fabulous and strange hybrid machine, with vast, yellow wings of fabric and bamboo atop three delicate-looking bicycle wheels. Except for the large motorcycle engine behind the pilot's seat, the curved, sail-like wings and slender brace wires give the contraption a vaguely nautical look. Tomorrow, on Independence Day, they will unveil their flying machine to the world, including a delegation of some of the country's leading aeronautical experts. Today, however, Curtiss and his colleagues ready it for a final test run.

As planned, the members of the Aerial Experiment Association have rotated responsibility for each prototype that the team has created together. Today, Curtiss, who has masterminded this airplane, will serve as pilot.

The machine has been painstakingly handcrafted, blending proven features with a few untested innovations. Like its predecessor, this is a biplane; its top wings arch downward at the tips while the bottom set arcs gently upward. The effect, as one observer will note later, is reminiscent of a large, sideways parenthesis. Among the novel features, perhaps the most notable is the means of lateral control. It sports a pair of small, triangular flaps at the tips of each set of wings. The wing flaps are adjusted by means of a yoke that fits over the pilot's shoulders. The pilot can keep the airplane from rolling out of control simply by leaning as it banks on a turn. Now known as ailerons, they are a feature of almost every modern airplane.

This latest AEA machine also boasts improvements to the wings themselves. At the suggestion of Octave Chanute, the team has coated them with a paintlike formula—a mixture of paraffin, gasoline, turpentine, and yellow ochre—that both reduces air resistance and enhances the craft's visibility.

June Bug, as Bell has christened the flying machine, is powered by the Curtiss shop's largest motor: a 40-horsepower, 8-cylinder, air-cooled engine weighing nearly 200 pounds. A year earlier, Curtiss used a similar motor, built onto an elongated motorcycle frame, to career at an astonishing 136 miles per hour along a track in Ormond Beach, Florida, earning himself the title of "the fastest man on earth." Now the same engine model drives one large screw propeller located behind the wings of an audacious machine designed to thrust him not just faster than anyone, but into the sky.

The AEA has entered the *June Bug* into competition for the *Scientific American* Trophy, a highly publicized prize offered for the first airplane in America that can prove, before judges, its ability to remain airborne for one kilometer. Unfortunately, though, while Curtiss has been practicing intensively with the *June Bug* since its completion a week earlier, he has only once managed to keep it airborne for that long. Yet the AEA has already confidently announced that it will fly for that distance before a visiting delegation of judges in a demonstration tomorrow. On this, the final dress rehearsal before the official trial, the team urgently hopes to live up to the claim with a successful dry run.

Beneath a nearly cloudless sky, Curtiss straps himself into the pilot's seat. The engine roars and the airplane's propeller whirls behind his head. As the rest of the team backs away, the *June Bug* starts to roll along the racetrack and then, almost magically, lifts from the ground. But just a few hundred yards into the flight, Curtiss feels the machine shake. He struggles at the controls, but there is little he can do as an unexpected gust knocks the *June Bug* askew, causing it to tumble roughly to the ground.

Fearing for Curtiss's safety, the team runs across the field to his side. The shaken pilot emerges unscathed, but then a bleak realization quickly takes hold. The airplane is badly damaged, its left wing broken, its front control smashed, and one of its wheels twisted nearly in half. With the invited aeronautical delegation due the next day, the AEA's vaunted, one-and-only prototype now lies before them as a crumpled wreck. Silence betrays the group's disappointment and gloomy sense of defeat.

Selfridge is the first to voice the unspeakable, suggesting that they postpone the public flight and telephone the delegation in New York City immediately to urge them to delay their visit. Curtiss alone is unwilling to accept such a setback. He insists that they can rebuild the aircraft in time, turning to Kleckler—as much for moral support as in the hope that, given his technical assessment, he will concur.

Kleckler, like the others, is skeptical. They roll the damaged machine back to its tent and start to debate the matter in earnest. Curtiss ignores the heated discussion, beginning instead to work quietly and furiously to repair the damage. His determination is infectious. Soon all are working in a frenzy, racing to and from the shop to repair the *June Bug*'s broken parts. With help from Kleckler and from Curtiss employees at the shop, the AEA actually restores the aircraft, completing a week's worth of repairs in just twelve hours.

By day's end, the team is confident that the *June Bug* is once again functional, but they can't risk flying it untested. So, in the last light of a very long day, they once again roll the aircraft onto the racetrack. Despite some trepidation lest he cause further damage, Curtiss gets the machine aloft for a one-minute flight.

It is too dark to fly the full kilometer in the refurbished aircraft. Nonetheless, after a perfect landing, Curtiss greets his teammates on the racetrack with a broad smile. He can't account for it, he says, but the *June Bug* seems to fly better than ever before.

The impression is not just his imagination. As he realizes later, the group, faced with a shortage of material, gave the updated version a slightly smaller elevating rudder, which made the craft easier to handle in the air.

As they roll the *June Bug* back to its tent, the AEA members all tiredly agree they would feel better having had a smoother and more thorough day of testing. But there is little they can do about the situation now. The aeronautical delegation is en route by train to Hammondsport. And the esteemed visitors will pass judgment on the AEA's machine. On the eve of the official trial for the *Scientific American* Trophy, the group has little assurance that the *June Bug* can meet the challenge. But thanks to their last-minute efforts, they take some comfort that the aircraft is at least intact for its debut.

For months, the *Scientific American* Trophy, established by the monthly science magazine, has been a major subject of discussion in aviation circles, especially at the staid, wood-paneled headquarters of the Aero Club of America where the shiny and substantial prize physically resides. At the start of the year, not long after the formation of the AEA, Curtiss personally admired the trophy on a trip to the club on Forty-second Street in New York City. It stood prominently displayed in a glass cabinet—a large and elaborate sil-

ver sculpture of an airplane resembling Langley's aerodrome circling the earth above a pedestal ringed by flying horses.

To Curtiss, and to all AEA members, the trophy has come to
assume a significance far beyond its monetary worth: a tangible and
much coveted validation of their efforts. They badly want their
names engraved on the cup for posterity. In late June, when the
team sees how well its newest prototype flies, the first thought is to
notify *Scientific American* immediately that the Aerial Experiment
Association is ready to compete for the trophy.

Curtiss and his friends are so eager, in fact, that Selfridge telephones Charles Munn, publisher of *Scientific American* (and also
president of the Aero Club), on June 24, before the *June Bug* has
even flown the requisite distance in tests. On behalf of the group,
Selfridge proposes to make the official flight ten days later. He asks
Munn to send a delegation of judges to Hammondsport.

Unexpectedly, Munn is somewhat hesitant about the AEA proposal. In particular, he balks at the team's desire to make the
demonstration in Hammondsport. Despite Bell's reputation,
Munn knows relatively little of the AEA, and he is reluctant to send
a group of judges hundreds of miles to such a rural locale.

In retrospect, it seems clear that Munn had hoped the trophy
would lure the Wright brothers into a public display. When he had
first announced the new prize in the pages of *Scientific American* in
the fall of 1907, Munn wrote that he hoped the trophy would spur
innovation in the aviation field. But his editorial also complained
that, four years after word broke about their work at Kitty Hawk,
the Wright brothers had yet to publicly demonstrate their alleged
invention. The Wrights' secrecy, Munn said, raised doubts about
what they might have accomplished. Actually, Munn's editorial
was far more measured than some in this period. As one newspa-

per editorial had already challenged, for instance, the Wrights "were either fliers or liars" and without proof it was beginning to look like the latter.

Munn had followed his goading editorial with a personal letter to Orville Wright, encouraging the brothers to try for the trophy. When he received word of the AEA's proposed flight, Munn wrote to Orville again the next day, offering to delay Curtiss and the AEA if the Wrights would consent to make a trial.

Orville and Wilbur remained unmoved. As with so many previous offers, the trophy failed to draw the Wrights into a public demonstration. In his reply to Munn, Orville complained about *Scientific American*'s stipulation that the aircraft must take off from level ground under its own power—meaning, in essence, that the prospective machine must include wheels.

As Orville explained, "All of our machines have been designed for starting from a track." Actually, built on sled runners, the *Wright Flyer* required not only fifty feet of track but an accompanying derrick several stories tall to get airborne. The brothers would drop a half-ton metal weight from the top of the derrick which, attached to their airplane by a cable and pulleys, would push the craft forward with enough thrust to get it aloft by the track's end. Orville griped to Munn that pneumatic wheels did not seem like a "satisfactory" thing to include on a flying machine. "Personally," he wrote, "I think the flying machines of the future will start from tracks, or from [a] special apparatus."

Ultimately, Munn's reluctance to deal with the AEA is no match for the enthusiasm in Hammondsport. Selfridge and Curtiss make a quick trip to New York City to appeal personally to him and other members of the Aero Club. They point out that the rules specifically allow a contestant to choose the trial's location. They know their chances of success will be far greater if they make the flight from familiar terrain, but they are also eager for the home-court advantage offered by the encouragement of friends and relatives.

After several hours of discussion, aided by support from secretary Augustus Post, the Aero Club agrees to send a team of observers and judges to Hammondsport. Still, most members of the visiting delegation have low expectations, as do the veteran reporters following the story who venture forth from New York City. They grumble about the three-hundred-mile trip and voice skepticism that the little-known AEA team will succeed.

For the AEA, the path to a working airplane has been remarkably swift, with surprisingly few missteps.

By December 1907, the group completed its first glider, the *Cygnet,* based on the unusual tetrahedral glider design Bell favored. The *Cygnet* was a fifty-foot-wide array of more than three thousand fabric-covered tetrahedral cells set into an aluminum frame. Bell developed the design, which looked like an oversized, triangular honeycomb, on the sensible theory that the small, pyramidlike openings would provide lift, even at relatively low speeds. On December 6, 1907, the team put the *Cygnet* on floats and, attaching it by a long cable, towed it behind the lake steamer *Blue Hill* on Nova Scotia's pristine Bras d'Or Lake. The *Cygnet,* pulled behind the boat, carried the daring Lieutenant Selfridge 168 feet into the air and remained aloft above the lake for seven minutes. In the excitement, though, the boat crew forgot to cut the line to the glider. As a result, when the steamer slowed, the *Cygnet* landed gracefully on the lake only to be dragged roughly through the water until it tipped over and broke up. Selfridge dove clear just in time to escape injury, but the *Cygnet,* painstakingly constructed over many weeks, was beyond repair.

As the AEA members had agreed, Selfridge oversaw the next attempt. After much discussion, he helped nudge the group away from Bell's complex design to experiment with the comparatively simple biplane glider long championed by Octave Chanute. The *Cygnet* showed promise, he argued, but it was cumbersome. With a unanimous desire to "get into the air" as quickly as possible, the group returned to first aeronautical principles for two productive months of experimentation with gliders modeled after Chanute's biplanes.

From the start, Selfridge wrote, the object of the AEA was "to walk in the footprints of those who had gone before and then advance beyond." Selfridge solicited advice from a wide range of colleagues, including everyone from William Avery, Chanute's assistant, to the Wrights themselves. In a letter to the brothers in January 1908, Selfridge asked if they might share some particular information about their understanding of where the center of pressure fell on a wing. Wilbur's reply, offering some advice and referring Selfridge to several other sources, would later be used by the Wrights as evidence that the AEA stole their invention, but the charge is all but baseless. While helpful, Wilbur's letter did little but confirm what the group was quickly learning on its own.

In the effort to glean as much extant knowledge as possible, the AEA was aided by its decision to move from Nova Scotia to Curtiss's Hammondsport shop. Not only could the group take advantage of the milder climate; they were stimulated by the large number of aviation researchers who had gravitated to the town. With the lure of Curtiss's motors, Bell exclaimed, no other place on earth could boast "such an assemblage of genius along the line of aerial work as Hammondsport."

Captain Baldwin, at work building his latest dirigible for the U.S. Army, had moved his entire operation to Hammondsport. Another aeronaut, Charles Oliver Jones, a former engraver, was building a dirigible powered by an eight-cylinder Curtiss engine. Lieutenant Alexander L. Pfitzner, a former Hungarian officer from Budapest, was on the scene attempting to build a lightweight monoplane, again around a Curtiss engine. And J. Newton Williams, a former typewriter manufacturer from Derby, Connecticut, had relocated to Hammondsport to build an experimental helicopter that would, eventually, lift itself several feet off the ground.

In addition, Augustus Post, secretary of the Aero Club and life-long ally of Curtiss and the AEA, came to town upon hearing of their work and, for several months, became almost an adjunct member of the team. An independently wealthy balloonist, Post was an enthusiastic observer and a cheerful extra hand. He was also a pro-lific writer who would ultimately pen some of the most colorful and detailed remembrances of Curtiss's exploits.

For all the ferment and camaraderie, though, the AEA operated in virtually uncharted territory. Everything the five men undertook had to be carefully thought out and crafted from scratch. The group needed to develop a thorough working knowledge of aerodynam-ics. But even more, with the help of Curtiss—and often with addi-tional help from workers in his shop—the AEA needed to master the finer points of airplane construction, from welding metal and laminating wood to sewing fabric and crafting fasteners that could withstand vibrations without coming loose. They experimented with a wide range of materials and suppliers, sending away as far as to Japan for bamboo. At the time, there were no ready supply houses for many of the materials they required.

By early March 1908, Selfridge's biplane, with a rudder on the tail and a single-plane elevator on the front, was ready to be tested. The group called it *Red Wing* because they had used the same red silk that had covered the *Cygnet*. Given the time of year, they decided to mount the plane on sled runners and try to take off from ice-covered Lake Keuka.

As the craft neared completion, the U.S. Army called upon Self-ridge—still a commissioned officer—to report to Washington, D.C., but despite his absence, the group moved ahead, fearing that the onset of milder weather could weaken the ice on the lake. They chose Casey Baldwin as pilot simply because, aside from the older Bell, he was the only one without skates and would thus be of little assistance on the ground crew. On March 12, 1908, before a handful of friends and acquaintances, *Red Wing*, the AEA's first motor-powered biplane, rose to a height of about ten feet off the ground and flew for about thirty yards before its tail buckled, forcing Baldwin to land. The aircraft had accomplished no more than a large hop, but the group was elated: their flying machine had risen into the air on its own power and under the pilot's control.

The AEA knew it was now on a promising path indeed. The team immediately wired the news to Selfridge, but detained in Washington, he would never see *Red Wing* fly. Five days later, upended from the side by a gust of wind, it crashed onto the ice. Pilot Baldwin escaped injury, but the accident destroyed the craft and its motor. It was a setback, but the AEA, hot on an inventive streak, hardly broke stride. The team was too excited by the chance to refine their design.

This time, it was Casey Baldwin's chance to oversee the work as the AEA hit a new level of intensity as a design team. In Hammond-sport, the group held discussions in Curtiss's shop almost every evening. In the shop's annex, which they dubbed the "thinkorium," they talked about everything: airfoils, atmospheric pressure, engine refinements, landing gear. McCurdy and Selfridge regularly faced off in a running debate over how best to improve a propeller's torque—the force with which it could move the aircraft. Whenever this subject came up, Curtiss recalled later, the team knew the argument was likely "to keep up until one or the other would fall asleep."

Bell's influence was clearly visible in the disciplined and formal procedures the AEA adopted. Each night the minutes of the previous meeting would be read and discussed, with notes assiduously kept by Selfridge. Because members, especially Bell, often had to leave Hammondsport to attend to other business, the AEA created a weekly, typed publication to chronicle their day-to-day progress. This internal newsletter, the *AEA Bulletin,* stands as one of the most remarkable firsthand records of technological development ever produced.

A major design problem confronted the group, especially after the *Red Wing* accident: how to control their aircraft's side-to-side motion. Indeed, lateral control was on the minds of aviation researchers around the world. As on other aeronautical matters, the AEA members showed not only prodigious inventive powers but also the strength they drew from the diversity of the team Bell had drawn together. They came up with the solution that would stand the test of time: *ailerons,* from the French for "little wings."

Aviation historians have long debated the provenance of the idea for these stabilizing flaps. The question is of particular interest because the matter would soon stand at the heart of the bitter lawsuit brought against Curtiss by the Wright brothers. In fact, neither Curtiss nor the Wrights can claim all the credit for the invention of ailerons. It is now clear that they had been imagined in full some forty years earlier and patented by a British inventor, M. P. W. Boulton. In 1868, Boulton spelled out the invention in detail, including the need for the flaps to tilt in opposite directions on each of an airplane's wings to keep the aircraft laterally stable.

While Boulton's prescient invention was fully functional, the same cannot be said for the airplane he designed, which never flew. As a result, his idea lay dormant until 1904, when a French aviation pioneer, Robert Esnault-Pelterie, experimented with two tilting, horizontal rudders in one of his glider designs. Brazilian aviator Alberto Santos-Dumont adapted much the same idea in 1906, adding two tilting, octagonal flaps into an early airplane he built *Number 14-bis* (so named because it was raised into the air for launching by his balloon *Number 14*). All these efforts predate the AEA's use of flaps, and yet none produced successful results in the air for their creators.

Nor did the AEA know of these efforts in 1907.* Regardless of the rich, overlapping milieu in which many researchers sought to solve the problem of lateral control, all evidence, including Bell's testimony, points to his having arrived independently at the idea for movable surfaces at the tips of the wing. As a seasoned inventor of great integrity, Bell was meticulous about assigning credit for his ideas. His habit was to write all his ideas in a notebook and have them promptly witnessed and notarized. He surely would have attributed the idea of ailerons if he believed he owed anyone such a

---

*As with so many other areas of aviation history, this is a matter of some disagreement. British aviation historian Charles H. Gibbs-Smith, tracing the intellectual history of ailerons, suggests that the AEA could have learned of them from a published account of Esnault-Pelterie's experiments with them in the January 1905 issue the French aviation journal *L'Aerophile*.

debt. As he would testify later, though, the idea for movable surfaces at the tips of the wing occurred to him from studying birds.

Certainly, the AEA knew a great deal about the problem of lateral
stability by 1907. And they were familiar with the Wrights' wing-
warping technique, as well as the other details of the Wright brothers' patent for their flying machine, which had been issued in May
1906. For example, Bell marveled in a letter to Mabel as early as
June 1906 that the Wright patent—which was, technically at least, a
public document that spelled out their invention—was receiving so
little notice in the United States.

But, just as the early uses of ailerons indicate, the issue of lateral
control had long been a secondary concern among those who strove
to create an airplane. It remains a controversial point, but there is
evidence that even the Wrights' vaunted concept of wing warping
had been in the aviation literature for many years. The early aviation
pioneer Otto Lilienthal experimented with wing warping in a glider
design as early as 1895. And, as Octave Chanute would later claim
in court, the naturalist and aviation expert Louis-Pierre Mouillard
had even patented a wing-warping method in 1898.

One way the Wrights' patent was influential, though, is that the
AEA took pains to steer clear of the Wrights' idea of bending
the wings of their airplane. Aware of the Wrights' proprietary claim,
the AEA looked for a separate method to keep their aircraft under
lateral control. Bell in particular tackled the problem with his usual
acumen. On March 20, 1908, he wrote to Baldwin from Washington, D.C., mentioning the aileron idea after mulling over the reports
the team had sent him about *Red Wing*'s demise. In this letter, Bell
suggests the tilting flaps, as well as several other related approaches
the group might pursue. "It may be that a lengthening of one wing
and a shortening of the other is what is wanted," Bell surmises, out-

lining a possible design in which an extension piece on the tip of a wing might open and shut like a fan. "This kind of action is employed by birds," Bell writes. "I have often seen birds suddenly reef their wings, so to speak, during a sudden squall, thus diminishing the supporting area of their wings." AEA's goal, as Bell puts it, should be to find a way to "reef one wing and expand the other."

The AEA quickly incorporated Bell's novel suggestion for ailerons into the airplane whose design Baldwin oversaw. With their supply of red silk depleted, they called the craft *White Wing*, covering its wings with cotton sailcloth. The aircraft would mark a pivotal step for the AEA, including many innovations—not just the ailerons but a stronger, laminated wood propeller and the inclusion of a wheeled, tricycle undercarriage.

On May 18, 1908, *White Wing* flew for the first time, covering a distance of 85 meters with Casey in the pilot's seat. On May 19, Lieutenant Selfridge became the first member of the U.S. Army to pilot an aircraft. Two days later, on his thirtieth birthday, Curtiss flew an airplane for the first time, setting a distance record for the group of 310 meters. Curtiss said afterward that the biggest surprise was how sensitive the plane was in the air. When he pulled back on the steering wheel, he was surprised at the way *White Wing* rose swiftly into the air. He immediately countered the effect so hard that the plane bounced to the ground. "As is usual in any balancing act," Curtiss said later, "the novice overdoes matters. . . . I realized that vertical control was a very delicate thing, and although I did my best to keep on a constant level, there was more or less hitching up and down through the entire distance."

White Wing was the cause of much elation and pride on the part of its designers. But like its fragile predecessors, the aircraft would not last long. It made just seven flights before crashing beyond repair with McCurdy at the controls, although he was not badly injured.

With the destruction of *White Wing,* the next AEA design fell to Curtiss and afforded the group the chance to consolidate all it had learned so far. During the first three attempts, Curtiss had mostly confined his input to the aircraft's power plant and propellers. While the long hours and constant aeronautical discussion had drawn the team close, Curtiss was always cognizant that he was the only member without an advanced degree and so he deferred, mostly watching, studying, and learning. But he was a quick study and a remarkable hands-on engineer, improving upon past failings, learning from prior mistakes. Now he had his chance to prove it.

As Curtiss and the group worked on the plane they would eventually call the *June Bug,* the nightly conferences continued. One evening, a particularly productive discussion about the way air passes over a wing—or airfoil—led Curtiss to surprise the group by designing and building an innovative wind tunnel. A coffin-like box, it had an electric exhaust fan on one end and a radiator pulled from an automobile on the other. As Curtiss demonstrated to the group, a model wing or aircraft could be hung inside the box and observed through the glass window in the top as a swift current of air swept through the apparatus. According to one account, Curtiss, in a fit of inspiration, had Harry Genung come and puff his pipe in front of the radiator to help the group actually see the air turbulence. As the smoke was sucked into the machine, the five AEA members watched fascinated as it flowed over the little wings, making swirls and eddies around the edges. Bell called it an enormous contribution to aeronautics, and all agreed that it helped to streamline their design. They were so taken with the contraption that they failed to realize that Genung was puffing so hard on their behalf he was making himself sick and had to be escorted outside for fresh air.

Genung's voluminous smoke would be replaced by threads of scarlet silk that would serve the same function of highlighting air turbulence for the increasingly scientific AEA designers. And, with the help of the wind tunnel and the group's lengthy, increasingly knowledgeable discussions, Curtiss introduced several vital modifications to the earlier AEA efforts. He lengthened the airplane's body to improve its horizontal, front-to-rear stability. For easier storage and transportation, he devised a means to make the wings removable and to allow the tail section to be folded up. And he refined the ailerons.

In just one month, Curtiss led the group to once again refine their design. The result, the *June Bug,* finally met their elusive goal: in an extraordinarily short and intensive design period, the AEA had built a fully controllable flying machine.

On July 4, the early morning train from New York City brings the distinguished aeronautical delegation to the tiny Hammondsport train station. The group includes nearly two dozen members of the Aero Club of America, among them Stanley Y. Beach, editor of *Scientific American.* Allan R. Hawley, a Wall Street stockbroker and balloon hobbyist, officially represents the club. None other than Charles Manly, Langley's former assistant at the Smithsonian, will serve as the official starter for the test. Other members of the delegation include Augustus Post; Simon Lake, inventor of the submarine; Karl Dienstbach, representing the imperial German government; George H. Gary of the New York Society of Engineers; Ernest L. Jones, editor of the *American Journal of Aeronau-*

tics; and Wilbur R. Kimball of the Aeronautical Society of New York. As this is the first publicly advertised flight in America, a large number of reporters, photographers, and even a motion picture crew have made the long trip from New York and other parts of the country to Hammondsport.

By 5 A.M., the rural roads are already clogged with traffic. Spectators find spots along the surrounding hills, where they have a clear view of Harry Champlin's racetrack. By midmorning, at least a thousand spectators have crowded in for the show; everything else in town has ground to a halt. Many bring picnic baskets, and an air of excitement and high spirits pervades the scene. As the morning wears on, families chat with one another, children frolic in the fields, and farmers stand side by side with the formally dressed visitors from out of town.

The weather, however, looks increasingly ominous. Although the day is warm, the wind is strong, and the threat of rain grows as the morning edges on. Curtiss, utterly immune to the crowd's growing impatience, simply will not fly until the skies settle to his satisfaction. Too much is at stake, he says, to risk another accident.

Early in the afternoon, a summer shower drenches the festivities, yet, as spectators huddle under umbrellas and blankets, their numbers still grow. Despite the miserable weather, few are willing to pass up their first-ever opportunity to see an airplane fly. In an effort to keep up the spirits of the eminent visitors and reporters, vintner Harry Champlin invites the dignitaries to duck inside his nearby winery during the rainstorm, offering them free food and his company's Great Western champagne. According to one report, Champlin later explained his generosity by saying that he wanted to help so that if Curtiss's flight failed, the reporters wouldn't treat him as badly as they had Langley in 1903.

Finally, with the approach of evening and the appearance of patches of blue sky, Henry Kleckler and several AEA members roll the *June Bug* out of its tent, where it has waited shrouded in mystery. Hawley and Manly promptly slog through the mud and wet grass to measure off the one-kilometer course, marking it officially and decisively with a pole topped by a red flag.

Upon their return, Curtiss starts the *June Bug*'s engine and quickly takes his place at the controls. With a wave of his hand, the rest of the crew lets go of the wings and steps back as their remarkable craft begins to roll along the muddy runway. Before a silent, awestruck crowd, the airplane rises into the air. And then rises further. As Curtiss struggles at the controls, something seems to be amiss. Continuing its steep ascent, the *June Bug* is now more than two hundred feet above the crowd, causing Lena Curtiss to loudly cry out, "Oh, why does he go so high? Do you think he's going to make it?"

Just then, forced to kill the engine, Curtiss glides to a gentle landing less than halfway to the distant marker. Some in the delegation, most notably Stanley Beach, are heard to mutter and scoff. But, undaunted, Curtiss and crew drag the *June Bug* back to the starting point and huddle for a conference. After a short discussion, Selfridge discovers the tail section has been accidentally set askew, tilted in a negative angle, causing the added lift. Making the minor adjustment, Curtiss once again climbs into the pilot's seat.

It is now around 7:30 P.M., but in midsummer there is still a good hour before sundown. On its second attempt, the *June Bug* once again bumps over the muddy ground and rises cleanly to an altitude of about twenty feet. Curtiss and craft are off and running.

Although Alexander and Mabel Bell returned to Beinn Bhreagh several days before, their daughter and son-in-law, Daisy and David

Fairchild, are among the witnesses that day and they offer a memorable recollection of the historic event. For David Fairchild the flight was "the experience of a lifetime." As he wrote shortly after the event, the people gathered around the aircraft suddenly backed away into the surrounding field. "Curtiss climbed into the seat in front of the yellow wings, the assistant turned over the narrow wooden propeller, there was a sharp, loud whirr and a cloud of dust and smoke as the blades of the propeller churned in the air.

"Then, before we realized what it was doing," Fairchild recalls, "it glided upward into the air and bore down upon us at the rate of 30 miles an hour. Nearer and nearer it came like a gigantic ocher-colored condor carrying its prey. Soon the thin, strong features of the man, his bare outstretched arms with hands on the steering wheel, his legs on the bar in front, riveted our attention. Hemmed in by bars and wires with a 40-horsepower engine exploding behind him leaving a trail of smoke and with a whirling propeller cutting the air 1,200 times a minute, he sailed with forty feet of outstretched wings twenty feet above our heads."

To Fairchild, and undoubtedly to many of the other spectators, the sight is overwhelming. All at once, he conjures up "strange visions of great fleets of airships crossing and recrossing both oceans with their thousands of passengers. In short we cast aside every pessimism and give our imaginations free rein as we stood watching the weird bowed outline pass by."

If David Fairchild's account captures the elation of witnessing the flight itself, Daisy Fairchild's recollections offer a sense of the pandemonium that ensues as Curtiss crosses the finish line. "As Mr. Curtiss flew over the red flag that marked the finish and way on toward the trees, I don't think any of us quite knew what we were doing. One lady was so absorbed as not to hear a coming train and was struck by the engine and had two ribs broke. . . . We all lost our heads and David shouted, and I cried, and everyone cheered and clapped, and engines tooted."

As the *June Bug* dips to a landing out of sight, the crowd spontaneously erupts in cheers and onlookers swarm over the potato fields, pastures, and adjacent vineyards and railroad tracks in celebration. As they reach Curtiss and his airplane just beyond a tangle of vines at the field's edge, they find him calmly examining the *June Bug*'s engine, though his enormous smile reveals his pride. Amid the turmoil, Hawley and Manly carefully measure the official distance of the flight, announcing to the crowd that Curtiss has flown 5,090 feet—just shy of a mile—or 1,810 feet more than the required kilometer.

Everyone jumps, cheers, and hugs one another. The shy Lena Curtiss joins in fully in the impromptu festivities along with the rest of the AEA members, as well as others from the Curtiss shop, including her good friends Harry and Martha Genung. Even the once-skeptical Aero Club members are beside themselves with delight to see the fulfillment of their fondest dreams. Some have studied and experimented with heavier-than-air flight for years. Now, at last, they have witnessed a self-powered flying machine carry a pilot into the air. One formerly skeptical newspaper photographer admits that, while his paper has had him chasing alleged birdmen for two years, he never believed they could really get off the ground.

The flight resoundingly wins the *Scientific American* Trophy for Curtiss and the AEA. Despite the prior claims of flight by the Wright brothers, the contest committee of the Aero Club will also ultimately award Curtiss the nation's first-ever pilot license, ruling that, with the *June Bug*, Curtiss has made the first officially observed flight in America and properly deserves the honor.

Curtiss, like the eye of a storm, remains eerily calm at the conclusion of his triumphant flight. He is proud but the attention seems to leave him even stiffer and more tongue-tied than usual. When a reporter asks about the flight, Curtiss can speak only of his focus and determination, noting honestly, "I could hear nothing but the roar of the motor and I saw nothing except the course and the flag marking a distance of one kilometer."

If Curtiss can't find the words to capture the significance of the flight of the *June Bug,* there is no shortage of others willing to fill in on his behalf. To the many astonished spectators and reporters on hand, it seems as though the world is changing before their eyes. Headlines across the country blare news of "the first official test of an aeroplane ever made in America." The Associated Press reporter on the scene calls the flight a matter "of the utmost importance." As others write, Curtiss and the AEA have done more in one afternoon to boost public faith in the promise of aviation than the Wrights had done in the five years since they claimed to have flown above the sand dunes at Kitty Hawk.

Reflecting on the day much later, Curtiss biographer C. R. Roseberry would note: "There was nothing he could do about it. The newswires had suddenly made Glenn H. Curtiss a famous man. Neither of the Wright brothers had yet flown in public. More than any other aircraft up to that moment, the *June Bug* convinced the world of the reality of human flight."

SEVEN

SKY KING

*Whoever will be master of the sky will be master of
the world.*

—CLEMENT ADER,
FRENCH AVIATION PIONEER, CIRCA 1909

Early in August 1909, eight months
into what is already an extraordinary year, Glenn Curtiss gazes
down at the crowd from the deck of the steamship *La Savoie,*
departing New York City for France. Curtiss's wife, Lena, and his
best friends, Harry and Martha Genung, stand among the well-
wishers on the dock below waving hats and handkerchiefs.

On deck, Curtiss sports a natty, wide-brimmed Panama hat and
a small goatee he has grown to cover a lingering scar he received
months earlier in an iceboat accident at Alexander Graham Bell's
estate in Nova Scotia. The getup almost succeeds in giving Curtiss
a suave, worldly air. But it can't mask the earnest Hammondsport
native underneath or hide his trepidation about the voyage ahead.

Curtiss is heading to Rheims, France, to represent the United

States at the world's first international flying tournament. He has never traveled to Europe—a considerably bigger trip even than to Bell's estate in Nova Scotia. He has never been on an ocean liner. Of more importance, the airplane lodged safely in the cargo hold belowdecks, has never been flight-tested. And yet now, at age thirty-one, with a career already full of show-stopping feats, Curtiss's participation in the *Grande Semaine d'Aviation* promises to mark a new international pinnacle.

Just over a year since Curtiss's *June Bug* flight, a vast sky of opportunity has opened to him—and the world of aviation has changed dramatically. Prodded by the success of the Aerial Experiment Association and by burgeoning developments in France, the Wright brothers finally decided to demonstrate their aircraft to the public in August and September of 1908. And even aside from the Wrights, an astonishing flurry of activity has thrust the airplane onto the European scene with a dynamism and pace of development few could have anticipated.

As one eminent aviation historian observes, 1908 marks the airplane's *annus mirabilis*—its miracle year. With the maturation of the internal combustion engine and a working understanding of the aerodynamics of lateral control, the final obstacles to manned, powered flight have been overcome. And almost all at once an assortment of airplane designs has begun to appear in Europe—primarily in France—with dozens of daring and glamorous aviators taking to the skies.

For Curtiss, the past year has brought both tragedy and opportunity. At the Wrights' first public demonstration in the United States—for the U.S. Army at Fort Myer in Virginia in September 1908—Orville crashed his *Wright Flyer* with Lieutenant Selfridge on board as his passenger. The Army had stipulated to the Wrights that they were interested in an airplane that could carry a passenger. Orville broke several of his ribs, but young Tom Selfridge, dedicated AEA member and undoubtedly the most competent and knowledgeable aviator in the U.S. military, became the airplane's first fatality. Selfridge's death was a terrible personal blow to Curtiss, and it also hastened the demise of the AEA.

With Selfridge's untimely death, the AEA's collective undertaking seemed to come to a natural, if heartrending end. The year-long contract the AEA members had made with one another was due to expire and they had achieved success far beyond their wildest expectations. While the remaining members remained close for the rest of their lives, the loss of a vital team member precipitated their somber decision to dissolve their formal association. McCurdy and Casey Baldwin chose to continue their aeronautical research in Canada with Bell as a senior consultant. Curtiss, with the group's blessing, opted to try manufacturing more airplanes in Hammondsport, building upon the success of the *June Bug*. In a somewhat impetuous move, Curtiss also decided to team up with a quixotic Aero Club member named Augustus Herring.

It is not entirely clear what Curtiss saw in Herring, but most likely he was taken with Herring's indisputably vast aviation experience as well as his claim to having numerous seminal aeronautical patents. Herring was urbane, educated, and as Curtiss put it admiringly, "a marvelous talker." Plus, he seemed to have an uncanny knack for attaching himself to many of the legendary names in the field of aviation. He had assisted Octave Chanute with his biplane glider design and had worked for Hiram Maxim in Britain when Maxim was attempting to build a flying machine toward the end of the 1880s. He had even briefly helped Langley with the aerodrome's construction. And he had won a contract from the U.S.

Army to build an airplane, even though he had yet to actually deliver it. To Curtiss, still a relative neophyte in the aviation field, Herring's résumé, contacts, and grandiose testaments of his accomplishments must have seemed like an important asset to a new company hoping to build and sell airplanes.

Before this new partnership began, Curtiss and Kleckler built an airplane under a commission from the Aeronautical Society in New York. Called the *Gold Bug,* it was the first commercially sold airplane in the United States—a fact that sent the Wrights into paroxysms of fury. Adding to the Wrights' indignation, Curtiss piloted the *Gold Bug* to victory in a second competition for the *Scientific American* Trophy. In this round, Curtiss flew nonstop for 24.7 miles over a circular course on Long Island, New York, close to double the 25-kilometer circle required to engrave his name for the second consecutive time on the nation's most prestigious, independent aviation award. Equally important, Curtiss more than amply demonstrated the airplane's capabilities to its proud new owners at the Aeronautical Society. They had little doubt that Curtiss had built—and they had purchased—by far the most successful flying machine in America.

On board the ocean liner *Savoie,* Curtiss sits on a deck chair, flanked by the two assistants he is bringing with him to Rheims. He has left his Hammondsport business in the trusted hands of Kleckler and Genung. For the trip, he has done his best to replicate the pair who have always offered him their unstinting support. To aid with the engine in Kleckler's stead, he has brought Tod Shriver, a promising young mechanic; while for the personal loyalty and all-around support usually provided by Genung, he has asked Ward

Fisher to come to France with him. A Hammondsport local, Fisher has been a good friend ever since Curtiss's bicycle-racing days. The three peruse an advance program advertising the meet and marvel at the turn of events that has brought them to this ocean crossing.

Just two months earlier, Cortlandt Bishop, the wealthy, debonair president of the Aero Club, heard of Curtiss's feat in the *Gold Bug* and cabled urging him to enter the Rheims meet. Curtiss hesitated. He had no airplane and had never tried to build one to race for speed. Bishop convinced him to try nonetheless. If for no other reason, Bishop said, Curtiss should enter the meet to represent the United States, especially because the Wrights, as usual, had declined to participate. Once again, the Wrights had protested to the Rheims organizers that the rules required contestants to take off on level ground on their airplane's own power without benefit of tracks or derricks. This time, however, the French organizers called the Wrights' bluff, offering to modify the rules on their behalf. But the brothers still refused to enter, even though Orville was scheduled to be in Europe at the time. Their petty objections assuaged, the Wrights were left only to gripe about the Rheims meet itself, telling the press that "circus performances like this do no good for the science of aviation."

For his part, Curtiss was captivated by Bishop's suggestion, but bringing a plane to Europe was a big undertaking. Bishop, who had inherited millions from his father's real estate interests, finally convinced Curtiss by offering to personally reimburse all his expenses if the plane failed to win any of the meet's $40,000 in prize money. Ever the businessman, Curtiss easily figured that he had nothing to lose. At the very least, he would see what Europe was like, and with luck, it could help attract a world of business to his new aeroplane manufacturing venture.

Although Curtiss knew from Bell and others that European aviators were making great strides, he received stunning news just shortly before his departure: On July 25, 1909, Louis Bleriot triumphantly piloted a small monoplane across the English Channel. Curtiss's own 24.7-mile flight on Long Island earlier that month had been slightly longer than Bleriot's 22-mile journey from Calais, France, to the Cliffs of Dover, England. But the distance was all but irrelevant. Fully capturing the European imagination, Bleriot's feat had an impact incalculably greater.

Because of his last-minute decision to enter the Rheims meet, Curtiss opted to build a machine much like the *Gold Bug*. To increase its speed, he made the new machine a bit lighter and added a new water-cooled, 50-horsepower engine and a longer propeller. There was no time to build many spare parts, but Curtiss did manage to pack an extra propeller in case the original broke. His chief concern—and, he believed, the linchpin to any possibility of success against his competitors—was the motor, just as it had been with his motorcycle triumph years earlier. As a result, he fussed relentlessly to refine the engine design, working all night long with Kleckler and Shriver on the eve of his departure to get it to reliably develop the desired 50 horsepower on the test block. Given the time constraints, though, the new motor had only one day of bench testing. The team didn't even have time to give the aircraft a trial flight.

Working up to the very last moment, Curtiss nearly missed the train to New York. Lena, Harry, and Martha had to rush ahead to Hammondsport station to convince the conductor to hold the train while Curtiss and his workers hurriedly packed the engine and dashed to the station, where the crate was quickly shoved into a boxcar alongside the packing boxes containing the aircraft's frame.

•   •   •

Upon his arrival in France in late August, Curtiss finds the country agog over the airplane. The Rheims meet offers the biggest and best example: twenty-two fliers are slated to attend. They will bring airplanes representing no fewer than ten manufacturers, further testifying to the ferment that has taken place in aviation, especially over the past year. European inventors have made many important incremental steps toward a working airplane ever since the turn of the century, but the work has recently grown into a full-blown renaissance. Most of the aircraft are French—Voisins, Bleriots, Antoinettes, and Farmans—but three French pilots are scheduled to fly Wright aircraft. Even though the brothers will not attend, three of their airplanes, assembled under contract in France and modified to take off from wheels, have been entered. Curtiss and his *Rheims Racer,* as he has named his entry, will be the only fully American entry. Curtiss knows little about the French-designed planes, but he relishes the chance to go head-to-head in the air against the Wright brothers' airplanes.

The continent is so keyed up over its first competitive flying spectacle that, as one U.S. reporter puts it, the excitement is hard to convey to an American readership. "It is as though," he writes, New York were about to host a simultaneous combination of "the Vanderbilt Cup Race, the Futurity, a Yale-Harvard boat race, a championship series of ball games between the Giants and Chicago, and a municipal election."

Seeking to create the aura and grandeur of a world's fair, the French have spared no expense for the weeklong event. The Grand Marquis de Polignac has supervised the planning with the Aéro-Club de France, agreeing to oversee the various races. Eager to attach themselves to this latest, glamorous, and wildly popular sport of aviation, the region's major champagne producers, including Bollinger, Moët et Chandon, and Veuve Cliquot, have contributed handsome sums toward the site's preparation and offered much of the meet's lavish prize money.

For the site, the committee has chosen the plain of Betheny, on the outskirts of Rheims, eighty miles east of Paris. Steeped in history, the ancient cathedral city is the heart of France's champagne-producing region, as well as the site where French kings have been crowned for centuries. Now, in the heady thrall of the early twentieth century, the organizers hope Rheims might forever be associated with something equally glamorous but considerably more modern: the crowning of a sky king.

To accommodate the race, the organizers have built grandstands for 50,000 spectators modeled after those at the world's elite horseracing venues. They have even constructed a railroad line and new stations to transport visitors to the site. And they have designed the ten-square-kilometer flying field to include hangars, barber shops, florists, and a 600 seat restaurant with 50 cooks and 150 waiters. Of course, they also included ample space for a terrace bar where patrons could sip the world's finest champagne as they watch the show.

With aviation excitement gripping France after Bleriot's flight across the English Channel, the Rheims event could not be better timed. The organizers had hoped for some 250,000 spectators over the course of the week, but by the time the *Grande Semaine d'Aviation* is through, they will attract more than twice that number. As the opening draws near, eager visitors have reserved even the tiniest bedrooms in humble houses near the site and driven the prices of accommodations to unheard-of heights, with suites at Rheims hotels going for as much as $5,000 for the week.

In Paris, Curtiss is greeted at the station by the ever-cosmopolitan Cortlandt Bishop and immediately feels out of his depth. Bishop is amazed to discover that Curtiss and his team brought the entire aeroplane, packed in crates, along with them on the train as personal luggage. As Curtiss explains, the freight carrier would not guarantee timely delivery of the aircraft to Rheims. Most shocking to Bishop, as it will be to almost all the European aviation enthusiasts, is how compact Curtiss's machine is. Curtiss's entire aeroplane fit into a passenger compartment across from the one they sat in. Among Curtiss's innovations, he has manufactured the wings in sections so that they can be more easily shipped.

The entire team handily travels through Paris in two cabs to the Gare de l'Est for the final train ride to Rheims. Leaving Shriver and Fisher in charge of the luggage, Bishop takes Curtiss for a quick visit to James Gordon Bennett, sponsor of the grand prize for the fastest 20-kilometer flight.

Through Curtiss's ties with Bell and his association with the Aero Club in New York, he had met many wealthy socialites, but Bishop and Bennett are in an altogether different league. Bishop, second president of the Aero Club of America, is a full-time patron of the arts and racing sports, with a passion for ballooning and impeccably tailored clothing. James Gordon Bennett, who inherited control of the *New York Herald* from his father, is an infamous, irascible figure. Publisher of newspapers in New York, London, and Paris, he built his publishing empire through shamelessly self-promoting stunts like the "scoop of the century" of sending war correspondent Henry Stanley to the wilds of Africa to find explorer Dr. David Livingstone.

Like Bishop, Bennett has enjoyed a lifestyle of outsized grandeur. He spent summers in Newport, Rhode Island, where he introduced his confreres to the game of polo and dared one friend into riding a polo pony to the upstairs floor of the most exclusive men's club in town. He ran his operations primarily by cable from Paris because he has been all but run out of New York society. Among the incidents that led to his unofficial deportation, two were particularly infamous: careening naked at top speed around Manhattan in his horse-drawn carriage one drunken evening, and urinating in the fireplace at a party given by his fiancée's family.

When Bishop and Curtiss arrive at Bennett's well-appointed office, the publishing tycoon regards the aviator-mechanic from Hammondsport with a critical eye, noting his odd goatee and his wrinkled, store-bought suit. Lanky and gaunt, with a serious demeanor that verges on taciturn when he is nervous, Curtiss lacks the panache of most of the European fliers. Bennett greets Curtiss cordially, complimenting him on his fine sportsmanship in entering the meet. But he does little to hide his displeasure when he learns that Curtiss has brought his entire airplane in a train compartment on the way to Paris, and that his reserve equipment consists of one extra propeller. Unsure of what to make of the sole American contender for glory, Bennett just whistles and pulls at his waxed moustache.

Curtiss amiably ignores the skepticism of Bishop and Bennett. Yet the weight of his undertaking finally hits him full force when he reaches the race site, three miles north of Rheims, and finds his assigned hangar. It stands in a long row of new buildings housing the aircraft of the other contestants. Aviators, mechanics, and paying customers flock around unabashedly, eager to make his acquaintance.

Despite the steady companionship of his assistants and the avid interest of fellow aviators—many of whom, he is amazed to note, are conversant with all his exploits—Curtiss has never felt so conscious of his rural upbringing and lack of sophistication. As he walks down the line of hangars, he marvels at the opulence of the European fliers and their wealthy backers.

With their ground crews, equipment, and entire spare airplanes, Europe's greatest fliers, Latham, Bleriot, Farman, Delagrange, Cockburn, Lefebvre, Paulhan, Tissandier, the Comte de Lambert, and many others, remind Curtiss of knights, with their many attendants and royal sponsors. Louis Paulhan's attire adds to the impression of nobility; he favors wearing elaborate outfits sewn from brightly colored silks like those of a horse jockey. Hubert Latham has brought two fully assembled, batlike Antoinette airplanes, with wide-spreading wings and sleek, new, in-line engines. Latham fits the bill too. A dashing young man born into a wealthy family of ship owners, he has spent his adulthood hunting lions in Africa, exploring the Far East, and racing speedboats in France before discovering the airplane.

Gabriel Voisin is also on hand. Along with his brother Charles, Voisin has established the world's first made-to-order airplane manufacturing business, building any designs sought by his often fanciful customers, as well as experimenting with his own boxy creations and those of his brother. He has come to Rheims with an entire field kitchen and the professional cooks to staff it.

Bleriot, fresh from his world-renowned, channel-crossing triumph, requires a series of sheds to hold the five airplanes he has brought to Rheims. Gleaming most threateningly among the fleet is a big monoplane he commissioned especially for the event. Bleriot, engineer and businessman, has been interested in the airplane since

1900, when he tried unsuccessfully to make an "ornithopter"—a flying machine with flapping wings. He made his fortune selling head-
lights for the thriving automobile industry and has spent the past
two years so obsessed with flight that he has nearly bankrupted
himself. Since his channel flight, however, Bleriot's fortunes have
changed as wealthy promoters have recognized the potential of
enormous profits from his exploits. The squat, effusive Bleriot,
with a big bushy moustache and birdlike beak, has brought a staff of
six mechanics and a mind-boggling array of equipment and spare
parts. His team even has a cleverly designed hydraulic machine to
measure Bleriot's engines' horsepower so that, before starting out
on a flight, they can make sure they are running at peak efficiency.

Glenn Curtiss has two mechanics, one small airplane, and a
spare propeller.

Yet, if anything, the austerity and simplicity of his operation
endear him to the press and other aviators. As one aviation afi-
cionado puts it admiringly, "One of the most noticeable things
about Mr. Curtiss is his American coolness. He and his mechanics
do just what is necessary, and no more. His machine, like the way it
is handled, is extraordinarily neat. "

The small airplane Curtiss has brought, the *Rheims Racer,* is a
direct descendant of the *June Bug* but with considerable refinements. A biplane with a bamboo frame, its wings are khaki-colored
and stayed by fine, steel-stranded cables. For the sake of simplicity
and greater control, Curtiss and Kleckler have dispensed with the
bowed-wing design of the *June Bug.* And, to make it as fast as possible, Curtiss has shaved its weight to roughly seven hundred
pounds and increased the diameter of the propeller from six to
seven feet, a change that required lifting the frame higher above the
ground to provide adequate clearance. The steering wheel has also

been refined. Like its predecessors, it sits immediately in front of the pilot, serving as a rear steering-rudder when the wheel is turned in either direction; when pulled directly back, the steering wheel alters the inclination of the front elevating planes, giving ascending or descending control of the plane, a design pioneered by the Aerial Experiment Association that would stand the test of time.

Bishop has arranged lodging for Curtiss and his two assistants in the house of a local Catholic priest. But as it turns out, Curtiss, Shriver, and Fisher spend most of their days and some of their nights in the hangar. Bishop has also invited an illustrious assortment of visitors, including a bewildering array of earls and countesses not to mention their American Gilded-Age equivalents: Vanderbilts, Goulds, and Astors. Working with Shriver and Fisher to assemble the *Rheims Racer,* Curtiss greets them awkwardly as they gather respectfully outside to watch. They invariably marvel to one another at how small the plane is—comments that do little to bolster Curtiss's confidence about the upcoming race.

And yet, Curtiss's design is notable for its impressive ratio of power to its small size. As a result, it is the quickest machine of all those at Rheims to launch at takeoff. Much to the amazement of the other aviators, Curtiss sometimes manages to get his airplane aloft after rolling fewer than fifty yards.

As his one advantage, Curtiss had hoped to keep his 50-horsepower engine a secret, but Bleriot has learned of the powerful engine and ordered an eight-cylinder motor hurriedly built that is reputed to yield 80 horsepower. "When I learned of this," Curtiss confided later, "I believed that Bleriot had the trophy as good as clinched."

•   •   •

On Sunday, August 22, the Rheims meet gets off to an inauspicious start. The weather is awful for the first three days of the weeklong event: prolonged rain turns the field to mud. Even more alarming are the strong, gusty winds that preclude almost any possibility of flying. Most of the thousands of eager spectators are ill prepared for the inclement weather. Ladies in long chiffon dresses muddy their elegant satin shoes. Ankle-deep in some places, the mud swallows the tires of many arriving motorcars, forcing them in a humiliating, low-tech concession, to let teams of horses tow them out.

Yet even the bad weather cannot dampen the irrepressible excitement surrounding the prospect of competition among the world's top fliers and fastest aircraft. Over the course of the week, the crowds keep getting larger, with the added glamour of attendance by royalty and titled nobility. In addition to the aristocratic cachet the organizers have sought, they also manage to attract a sea of "regular folk," who buy cheaper admission tickets that allow them to watch the show from the open grounds. According to one observer on hand, the throng stretches literally for miles around the course, in some places as many as forty people deep.

The press, too, is caught up in the excitement. As one flowery newspaper dispatch to America crows: "Never since history began have there been witnessed such scenes of wonder . . . so presagent of a change in the life of man upon earth."

Armand Fallières, president of France, along with leading members of his government, are among the dignitaries from a score of countries attending the meet. The king of Belgium is in attendance. High-ranking military observers, like General John French of Great Britain, have come to see what flying machines can actually do, as has a delegation of engineers sent by the emperor of Japan. David Lloyd George, soon to be British prime minister, attends Rheims and is deeply impressed. "Flying machines are no longer toys and dreams," he says, "they are an established fact. The possibilities of this new system of locomotion are infinite."

Finally, toward the end of the first day, the weather clears just enough to fly. White flags appear on the distant hangars, alerting the public that a demonstration is about to begin. The three *Wright Flyers,* piloted by Paul Tissandier, the Comte de Lambert, and Eugene Lefebvre, take to the sky in a preliminary display.

Lefebvre ascends to three hundred feet, dives to within a few feet of the ground, then climbs while the Comte de Lambert flies beneath him. As a French correspondent puts it, the crowd is treated to a preview of "the wonders that the near future has in store for us."

Excitement builds as the week progresses. Many aviation records are broken. At one point, seven airplanes take to the skies together, an awesome sight for aviators and spectators alike. Curtiss, however, holds back. With only one plane to gamble, he resists the urgings of others to enter any of the other competitions before the most prestigious speed race.

Curtiss does, however, make a number of short practice flights to get familiar with his new airplane and make sure it is functioning properly. Even these forays underscore the risk involved. At the end of one of his test flights, he lands off the track and is thrown out of the plane as it veers into a field of tall grain. Spraining his ankle, Curtiss is forced for the rest of the meet to walk with a borrowed cane.

During breaks from their avid interest in the action in the skies, Curtiss and his mechanics work with one all-important goal in mind: to coax as much speed as they can from their airplane. The Wright brothers, however, have something quite different planned. Having striven arduously to secure patents all over the world, they now intend to control the field and collect royalties from all practitioners. And they decide to crack down on what they perceive as patent infringement at a time that will create maximum impact: just as the first international air meet heads into full swing. To their way of thinking, every airplane on the field at Rheims, except the three of their own models, infringes upon the brothers' patents.

Five days after Curtiss sails from New York, Orville Wright goes to Berlin for a series of demonstration flights, in an agreement with the just-organized German Wright Company. Angered at all the press Curtiss has been receiving, he writes Wilbur before leaving: "I think best plan is to start suit against Curtiss, Aeronautic Society, etc., at once. This will call attention of public to fact that the machine [*Gold Bug*] is an infringement of ours." And Wilbur responds: "If the suit is brought before the races are run at Rheims, the effect will be better than after." With Orville in Berlin, trying to close a deal with the German government, Wilbur has busied himself filing suit against Curtiss. As he explains to Orville: "Trophies are one thing, business another."

The Wrights' opening salvo, the day Curtiss first assembles his plane in Rheims, is to file suit against the Aeronautic Society as the buyer of Curtiss's *Gold Bug*. They seek a court order to stop the society from exhibiting the plane, asking for financial damages, and insisting that the plane be destroyed. Within a few days, the Wrights' lawyers also serve papers in Hammondsport on Mrs. Lena Curtiss and Lynn D. Masson, secretary-treasurer of Curtiss's firm, charging them with infringing their wing-warping patent. In Hammondsport, Judge Monroe Wheeler gives out a statement in his capacity as the Curtiss company's president and head counsel: "These suits will be defended, and it will be the policy of the defense to disprove all claims of infringement."

News of the Wrights' suit spreads quickly, shocking the fliers at Rheims. The overwhelming reaction of the European aviators is antagonistic to the Wrights for trying to establish a monopoly on flight. A patent fight is bad enough, but the timing seems particularly vindictive, considering that Curtiss is the sole U.S. representative in a world championship contest the Wrights have refused to enter. Many of the French fliers offer moral support to Curtiss; after all, they too are vulnerable to the Wrights' aggressive legal maneuvering. But, for now, Curtiss will bear the brunt of it.

The support of the French fliers bolsters Curtiss's confidence that the AEA's ailerons are a distinct and separate invention from the Wrights' technique of bending an airplane's wings. Ailerons operate completely separately from the airplane's rigid wings, tilting to create extra drag on one side of the airplane or the other to correct unwanted lateral tilting in flight. Far simpler and more efficient than the Wrights' system, which needs to be tied into the plane's rudder, ailerons are already becoming the industry standard. Reflecting the prevailing sentiment at the meet, Curtiss wires home not to worry. "No one here thinks there is any infringement," he writes to Lena, adding optimistically that the Wrights' suit might be a bluff.

Curtiss will learn soon enough that the Wrights are not bluffing. For the time being though, the clouds that have plagued the air meet finally begin to lift, even if the Wrights, in one dramatically timed move, have managed to hang dark psychological ones over Curtiss.

The weeklong meet reaches its climax on Saturday, August 28, with the race for the Gordon Bennett Trophy. Despite the Wrights' machinations, Curtiss, an intense competitor, has a race to win. He has an athlete's ability to shut out everything but the event at hand, born of long years racing bicycles and motorcycles. And in this case, he wills himself to put aside all thought of the Wrights' legal claims for the moment, letting the Gordon Bennett Trophy—without question, the most coveted prize of the meet—occupy his full attention.

Contenders for the $5,000 prize and trophy donated by the newspaper mogul must fly twice around the rectangular ten-kilometer course, staying to the outside of pylons several stories tall that mark the airfield's four corners. The way the race is organized, the contestants can make their official flight at any time between 10:00 A.M. and 5:30 P.M. Curtiss, who has always favored flying at the first light of day, decides to make his practice run at the earliest possible moment. At ten o'clock sharp, it is, finally, a perfect summer morning, clear and windless. In his brown leather jacket and visored cap, Curtiss takes his place in the pilot's seat. His lithe aircraft seems to almost hop into the air on takeoff, and the crowd watches his practice run intently.

Once Curtiss is aloft, his *Rheims Racer* jolts and rocks roughly through the course. The air, as Curtiss later describes it, is "boiling," with strong thermal updrafts caused by the now-hot summer sun beating on the open plain. The effect in flight is like repeatedly hitting vicious, invisible boulders while traveling in an automobile at high speed. Landing safely after the trial, Curtiss is unhurt, but he is badly shaken by the turbulent flight. Checking his time, though, Curtiss realizes that if he can power his way through the bumps, the windless morning is likely to offer the day's best conditions.

He decides to make his official flight for the trophy without delay.

This time, after takeoff, Curtiss stuns the sea of spectators by circling higher and higher to nearly 500 feet, then swooping into a steep dive to build as much speed as he can across the starting line. Full throttle into his official trial, Curtiss draws upon his experience as a motorcycle racer to lean sharply into the turns. He banks so steeply and shaves the pylons so closely, in fact, that many in the stands gasp, sure he will clip his wings. The stunned crowd has never seen anyone fly this way before, nor has Curtiss ever pushed an airplane so hard. All the while, he has to contend with turbulence that has grown even stronger since his practice run. The violent shocks lift him completely out of his seat. Curtiss has to wedge his feet tightly against the airplane's frame to keep from being thrown from the pilot's seat.

As Curtiss sweeps past the finish line and slows to a landing, a mob of cheering Americans rushes upon him. He has set a new world's record, flying the course in just under 16 minutes, with an average air speed of 46.5 miles per hour. As the first contestant to fly, however, Curtiss must now sit by and watch the other contenders, each time wondering if they will beat his time. It is a fate he likens later to that of "a prisoner awaiting the jury's verdict."

Over the course of the morning, there is much to see. The British flier George Cockburn runs into a haystack on his flight and fails to finish his trial. Latham makes his try in the afternoon, flying at low altitude, as he always prefers to do, ending the run with an average speed 5 miles per hour slower than Curtiss's. Most of the other contestants, like Farman, do not even come that close, averaging about 35 miles per hour at their top speeds.

Notably, most of the other entries easily outpace the Wright planes. Even in 1909, the fundamental limitations of their design are evident. Much the way a bicycle cannot maintain its balance unless it is moving, the Wrights have purposefully designed their planes to be inherently unstable, believing, mistakenly, that this is an essential factor to control in the air. As a result, the *Wright Flyers* are especially difficult to fly and require the pilots to actively control them in all three dimensions—pitch, roll, and yaw—during every moment they are airborne.

Bleriot, however, has yet to fly.

All day long, Bleriot tinkers with his assorted airplanes. He adjusts the engine on one and attaches a new propeller to another. Over the course of the day, he takes several practice laps but touches down each time, frowning, dissatisfied with some aspect of his aircraft.

By late afternoon, Bleriot has settled upon *Number 22*, his largest airplane and the one Curtiss fears most. It is equipped with the much-ballyhooed eight-cylinder, 80-horsepower motor. Equally worrisome to Curtiss, Bleriot fits the plane with a large, four-blade propeller.

As the afternoon progresses, Bleriot's delays cause mounting concern among the quarter million mostly French spectators assembled for the main event. Finally, at the last possible moment, Bleriot signals that he is ready. It is 5:10 P.M., just twenty minutes before the race's official close. The tension of a long day has begun to wear on Curtiss. When he sees Bleriot take off, he is sure the Frenchman will prevail. As Curtiss later recalls, "It looked to me as if he must be going twice as fast as my machine had flown."

Bleriot doesn't bank his large craft the way Curtiss had, but he makes two flawless laps of the course and lands to thundering applause from the mostly French crowd. As Bleriot walks over to the judges' booth, Curtiss keeps a respectful distance. The crowd is absolutely quiet until the silence is suddenly pierced by a joyous shout from near the judges' stand. It emanates from none other than Cortlandt Bishop. Flailing his arms, Bishop runs to Curtiss, shouting, "You win! You win!" As Bishop explains, Curtiss has beaten Bleriot's time by six seconds.

Almost before Curtiss can comprehend the news, the band strikes up the "Star-Spangled Banner" and an American flag is raised above the judges' stand. The predominantly French audience, stunned at first, graciously joins in the raucous applause of the American delegation to honor Curtiss's achievement. Newsreel footage shows a young American beaming at the apex of his career. Henry White, the U.S. ambassador to France, rushes from his box in the grandstand with his party—including Mrs. Theodore Roosevelt, wife of the ex-president (Teddy is away hunting big game in Africa) and three of the Roosevelt children, Ethel, Quentin, and Archie—to congratulate Curtiss in the name of the government and people of the United States. By Bishop's side, James Gordon Bennett swaggers over to shake Curtiss's hand, beaming as though he had believed in Curtiss all along.

Just then, Bleriot bursts into the midst of the patriotic scene. He flings his arms around the surprised and flustered Curtiss and kisses him heartily on both cheeks.

The close of the *Grande Semaine d'Aviation* throws France further into its frenzy of aeronautical excitement. As news of Curtiss's upset victory spreads, the always shy aviator is nearly overwhelmed by his sudden celebrity. He is the subject of adulation everywhere he goes. Kings, ministers, and wealthy socialites deluge him with offers to fly his plane all over Europe. Bishop calls his feat, "the greatest sporting victory the world has seen." Banner headlines proclaim him

"Champion Aviator of the World," and, recalling his past glory on his motorcycle, "Fastest Man of the Earth and Skies."

Christening its new hot-air balloon the *Curtiss No. 1,* the Aéro-Club de France urges him to ride as passenger on its maiden voyage along with Bleriot and Bishop. In his first ascent in a spherical balloon and basking in his good fortune, Curtiss marvels at the breathtaking French countryside below as Bishop's chauffeur follows them on the ground to drive them back to Paris at the end of the day.

As proud as he is for what he has accomplished, Curtiss would have preferred to go back to tinkering quietly with new improvements to his airplane. He is distraught by the Wrights' lawsuit and knows he must swiftly return to handle his company's affairs. Plus, he misses Lena; never again will he leave her behind on a major trip. But tonight he must attend a dinner in his honor at the U.S. embassy where five hundred elegantly attired people will dine in stately splendor. As the *New York World* accurately reports, Curtiss seems to "fear ceremony more than he fears the most perilous flight."

The guest of honor has little choice but to attend, but Curtiss does try to get Bishop to arrange it so no one will call upon him to make a speech. The effort is futile. The moment Curtiss walks into the grand dining room, arm in arm with Bleriot, he receives a standing ovation, and he remains the center of attention for the entire evening. Ambassador and Mrs. Henry White are particularly gentle and kind hosts. And Curtiss finds neither the gold-plate service nor the lavish repast as intimidating as he has imagined. But, as he recalls later, "when they wanted me to stand up and make a speech, I was lost." Nonetheless, the audience is forgiving. Curtiss has earned their admiration the hard way: in the sky. And they are willing to crown him the master of it, tongue-tied or not.

To make it official, his latest accolade, the Gordon Bennett Trophy, sits prominently in the middle of the table of honor. Curtiss finally gets a chance to inspect it for himself. It is a large, silver sculpture that fittingly depicts a person climbing a mountain and reaching upward toward an intricately detailed airplane. It is a handsome award, but Curtiss notices at once its fundamental irony: the hiker is reaching toward a precise and unmistakable rendering of a *Wright Flyer*.

PART III

WARPED WINGS

EIGHT

GROUNDED

*Mr. Curtiss's opinion, as expressed to us at Rheims, is
that the proprietors of the Wright patents think no one
can make an aeroplane without them and, said he,
"they're about the only people who do think so."*
—*NEW YORK DAILY TELEGRAPH,* 1909

Following his victory at Rheims, an assortment of Europe's royalty and wealthy socialites fete Curtiss for nearly two weeks and acclaim him an international hero. But for Curtiss nothing can match his homecoming in September 1909. On a cool, damp evening, as his train pulls into the Hammondsport station, Curtiss marvels wide-eyed as the skies over Lake Keuka explode with an extravagant fireworks display to mark his arrival.

As Curtiss makes his way among the jubilant crowd, bonfires blaze by the lake and the ground shakes with the boom of ceremonial cannon fire. A committee of admirers has erected a triumphal arch across from the station, with the word "Welcome" emblazoned in electric lights. The entire town is festooned with American flags.

Corks from New York champagne bottles pop into the air all around him and people raise glasses brimming with the local bubbly in tribute. And, of course, there is an outdoor stage in the center of town where it seems everyone Curtiss knows expects him to make a speech. Yet even here before the adoring assemblage of friends and neighbors—and with Lena at his side—words mostly fail him.

"Ladies and gentlemen," he says, visibly overcome by the festivities, "I'm back from France. Had a very nice time. Had a whole lot of luck and a little success. As you know," he continues haltingly, "I just had a common school education. I didn't learn any words big enough to show you my appreciation of this welcome home."

His words are quite big enough for Hammondsport and the town elders are especially proud. Their local son with his common school education has accomplished a remarkable feat on the world stage. Of his own initiative and after only an extraordinarily short period of experimentation, he designed and built an airplane that proved itself the fastest in the world. In France, when the editors of the magazine *L'Auto* assessed the aerodynamic efficiency of the different planes at the Rheims meet, they rated Curtiss's plane number one and ranked the *Wright Flyers* in last place.

But, for all the accolades and festivities, Curtiss can't shake a sense of foreboding. He may have mastered the engineering and technical aspects of the airplane, but he is far less confident about the legal quagmire looming before him.

Most pressing is the fiasco that has come of his alliance with Augustus Herring, which he had hoped would reap rich rewards in the young aviation field. Herring has failed to deliver the vital working capital he had promised in his partnership with Curtiss; even more worrisome, he has refused to reveal a single item of his vaunted inventions and patents. By now it seems painfully obvious to everyone in Curtiss's company that Herring is an unscrupulous fraud. As Curtiss will lament to Cortlandt Bishop, who has invested in the company, Herring "has fooled Mr. Chanute, Hiram Maxim, Professor Langley, and the U.S. government." Now, Curtiss writes, he has "deceived you and me."

Even as Curtiss tries to extricate himself from his business ties with Herring, he must contend with the much larger and more vexing problem of the Wrights' lawsuit. It seems inexplicable to Curtiss that his airplanes—with their rigid wings and ailerons—could be thought to infringe a patent that specifies the flexing of an airplane's wings in conjunction with the use of the rudder for lateral control. If the Wrights had conceived of ailerons themselves, Curtiss reasons, they would have explicitly incorporated the design into their patent.

Nonetheless, Curtiss knows that the technological complexity of the Wrights' case will make it difficult to successfully defend himself in court. And he cannot ignore the fact that the Wrights, represented by a top patent lawyer, seem determined to punish him for entering the field of aviation they had hoped to control alone.

Curtiss knows he needs the best advice he can find. Bell is encouraging and supportive, but, understandably, he is hopelessly partisan on the subject of ailerons. Meanwhile, Curtiss has agreed to demonstrate his airplane at an exhibition in St. Louis. Much ballyhooed, it will be the first demonstration west of the Mississippi of an airplane in flight. So after just a brief stay in Hammondsport, he takes Lena with him on the road. The flight is a tremendous success and, to celebrate, the two buy a big, new Chalmers-Detroit automobile to drive home. On the way, Curtiss arranges to visit Octave Chanute at his Chicago home. As perhaps the world's leading authority on aviation history, Chanute has intimate knowledge of the Wrights' aircraft and their process of invention leading up to it.

The visit with the venerable Chanute, now seventy-seven years of age, is both edifying and heartening. "I spent two evenings in Chicago with Dr. Chanute," Curtiss writes Bell confidently, "and from what he says the Wrights have little chance of winning."

Following Curtiss's visit, Octave Chanute issues his views publicly on the matter of *Wright v. Curtiss.* The Wrights, he says, are not the first to incorporate the concept of warping an aircraft's wings. "On the contrary," Chanute writes, many inventors have developed the notion since the time of Leonardo da Vinci. "Two or three," he says, "have actually accomplished short glides with that basic warping idea embodied in their machines."

Among these inventors, Chanute cites the work of his French colleague Jean-Pierre Mouillard. According to Chanute, in 1898 in Egypt, Mouillard successfully tested a glider that employed wing warping. With Chanute's help, Mouillard patented the device in the United States but never developed the idea further because he suffered a stroke and died shortly after its invention.

Furthermore, Chanute argues, the Wrights' design owes a heavy debt to his own work and that of many others. As he notes, "When the Wrights wanted to start, they wrote to me that they had read my book on gliding and asked if I would permit them to use the plans of my biplane." Chanute made all his data freely available to them. "I was glad," he writes, "that someone wanted to continue the work."

Chanute is clearly angered to see the Wrights now seeking to "shut off" the work of others. "Public competition," he writes emphatically, will best serve the field of aviation. This conviction offers Chanute's ultimate testament on the matter; he dies within the year.

Even if the patent office and the courts judge the Wrights' wing-warping patent to be a wholly novel invention, Curtiss feels buoyed by a growing body of analysis that considers ailerons to be outside the scope of the Wrights' patent. Most notably, Thomas A. Hill, a prominent New York patent attorney who studied the Wrights' patent on behalf of the Aeronautical Society, argues persuasively in the October 1909 issue of *Aeronautics,* the society's journal, that ailerons fall outside the Wright claims.

Curtiss's design, Hill explains, may produce an effect similar to that of the Wright brothers' airplane, but it only does so through the use of pivotally mounted "supplemental or auxiliary surfaces, planes or rudders." It is, he notes plainly, impossible to flex, warp, or otherwise bend or move the main surface on a Curtiss aircraft.

As Hill sees it, nothing in the Wright patent suggests "that they at any time intended to use supplemental surfaces for accomplishing substantially what they accomplish by warping, flexing, bending, twisting, or otherwise distorting the lateral margins of their main plane." The omission, he concludes, casts doubt on the Wrights' chances of success in their lawsuit.

The opinions of such distinguished experts as Hill and Chanute bolster Curtiss's confidence. But he soon learns one exceedingly disturbing fact: his case will be heard by a judge notorious for his predilection for expansive patent interpretations.

It is a very unlucky and fateful piece of news indeed.

Judge John Raymond Hazel, appointed to the federal bench in 1900, brought relatively little legal experience to the job. His nomination was widely criticized as political payback for his work on behalf of President McKinley's Republican Party. Almost immediately after his appointment, the controversial Hazel presided over one of the biggest patent battles of the day: a fight over the invention of the automobile. His handling of that lawsuit would earn him a good deal of notoriety.

In the case, George Selden, a lawyer and part-time inventor from Rochester, New York, had built a rudimentary three-cylinder engine in 1878. Selden never perfected the motor but he recognized the possibility that it might be used in a "horseless carriage"—a quest akin to the Holy Grail for many inventors near the turn of the century. Selden never actually built such a vehicle. He never even tried. He was, however, indisputably the first to apply for a U.S. patent on the notion of using a gasoline engine to propel a carriage—an invention that would, of course, come to be called an automobile.

He may not have been much of an inventor, but Selden was an accomplished patent lawyer and he managed to secure U.S. Patent No. 549,160 for his "road engine." Then, he cleverly nurtured along his patent in a "pending" status—a cunning but legal abuse of the rules of his day. Repeatedly amending his application slightly every two years as the rules generously allowed, he thereby dragged out his proprietary claim for more than a decade before it went into effect, waiting in a kind of legalistic ambush until someone actually did invent an automobile.

Sure enough, Selden's plan worked. He allowed his patent to come into force in 1895 after he saw actual automobile designs start to emerge on the horizon. Before long, he had forced nine fledgling automobile manufacturers to pay a royalty of 1.25 percent of the retail price of each automobile they sold for the right to use the

Selden patent. Casting aside the fact that countless inventors had foreseen the possibility of an automobile, Selden had the U.S. government's imprimatur. The idea for the automobile belonged to him, he believed, and he had the patent to prove it. By 1906, Selden was receiving royalties from almost every automobile maker in the United States.

Around this time, when Henry Ford set out to manufacture automobiles, he too was initially willing to pony up his royalty fee. But, branding him as a mere "assembler" of cars (and doubtless scared that he might undersell existing makers), Selden and the association he had set up to handle the patent refused to grant Ford a license, using the patent—as was technically their legal right—to block him from selling cars entirely. The move, needless to say, brought out Ford's fighting ire.

Ford went to court to challenge what he saw as not just a bad patent but a broken system. He lambasted Selden's claim, calling it "worthless as a patent and worthless as a device." Unfortunately for him, however, it was Judge Hazel who heard Ford's case. Undeterred by the vague and flimsy nature of Selden's claim, Hazel upheld Selden's patent, ruling that, as the first in its field, it deserved an expansive interpretation.

After a costly and well-publicized struggle that dragged on for years, the federal appeals court eventually overturned Hazel's verdict, ultimately agreeing with Ford's lawyer that Selden had disclosed "absolutely nothing" of social value in his patent. But the hard-fought case left Ford bitter despite the victory, with an abiding distrust of monopolistic power and, not surprisingly, more than a passing interest in Curtiss's plight.

Before Curtiss's ordeal was through, Ford would offer his assistance. As one story goes, while Curtiss was dining with his associate Lyman Seely one afternoon in the Brevoort Hotel in New York, a trim, white-haired man left his own party and came to Curtiss's table. He offered to help in any way he could in Curtiss's lawsuit, then walked back to his table almost before Curtiss had a chance to thank him.

Curtiss didn't know whether to be grateful or ashamed that his situation was so pathetic as to warrant the pity of complete strangers. While he never imagined he would pursue the matter, Curtiss asked Seely if he knew the man who had just offered his assistance.

"You're joking, aren't you?" Seely replied. "Surely you recognized Henry Ford."

In the matter of *Wright v. Curtiss,* Judge Hazel wastes little time. On January 3, 1910, he grants the Wright brothers the preliminary injunction they seek. The decision prohibits Curtiss from selling or exhibiting his airplanes. And it stuns almost every player in the youthful aviation field on both sides of the Atlantic.

For Curtiss it is a terrible blow. He learns of the injunction while readying his plane for the first international air meet in America— the "Air Tournament of Los Angeles"—beginning January 10, 1910. He recognizes immediately how bad the news is. Even though the case has yet to be tried on its merits, the judge has nonetheless deemed the Wrights claim strong enough to justify halting Curtiss's operation immediately.

On the eve of the Los Angeles event, the organizers worry whether Hazel's injunction might prohibit Curtiss's flight at an aviation meet. The press speculates that the Wrights might even try to shut the meet down entirely to ground the many "infringing" non-

Wright aircraft on the program. To mollify the nervous organizers, Curtiss explains that he will not, strictly speaking, be making an exhibition for money so the event should not pose a problem. He offers to post a bond pending his appeal of the injunction, and the organizers agree to let him participate.

Curtiss approaches the problem like the mechanically minded inventor he is. From his perspective, his airplanes' aileron system of lateral control has one difference from wing warping that is so glaring, it alone should turn the case in his favor. The way the Wrights' patent is written, their wing-warping system is tied into their airplane's rudder. When the wings are twisted, the Wright aircraft tends to turn. The Wrights correct for this tendency by wiring the rudder to turn slightly in the opposite direction whenever the wings are warped for lateral control.

Curtiss's aileron system is completely independent of the airplane's rudder. If he can simply demonstrate this, Curtiss figures, Judge Hazel will surely have to overturn the injunction. There at the meet in Los Angeles, Curtiss hatches a plan to publicly prove his point. And he announces it to the reporters assembled to cover the meet:

"In the arguments of their lawyers," Curtiss says, the Wrights have convinced Judge Hazel "that my machines depend on the vertical rudder to maintain equilibrium." In the Wrights' airplanes, he says, "the warping surface of the planes gives the machines a turning tendency which the rudder has to overcome." On the contrary, he says, "the rudders on the Curtiss machines have no such function." And, he says, he will prove his point with a demonstration.

Following the announcement, Curtiss, along with two other pilots flying Curtiss airplanes at the meet, set out to disprove the Wrights' claim. Immobilizing their airplanes' rudders, they each make straightaway flights before thousands of spectators. On each flight, the pilots move the ailerons vigorously back and forth, causing the airplanes to rock laterally like a boat buffeted broadside by strong waves. And on each flight, as corroborated by a host of expert observers, the planes display no turning tendency whatsoever from the use of the ailerons alone.

Much to Curtiss's chagrin, the spectacle does nothing to sway Judge Hazel. It does, however, become yet more fodder in the increasingly acrimonious case. According to the Wrights, Curtiss's claim is technically wrong and runs counter to their superior understanding of "the unchangeable laws of nature." Ailerons, they say, like wing warping, cause a turning effect that needs to be corrected by the airplane's rudder. In an affidavit, they charge that Curtiss's demonstration in Los Angeles only shows "his incompetence to give expert testimony as to what actually occurs on his machine, and even seems to raise a direct question of veracity."

Curtiss counters that his claim is based upon a fact witnessed by scores of knowledgeable observers. And, he adds tartly, "the actual facts are, I understand, the crux of the matter, and not the theories which may be advanced."

The truth, doubtless fully understood by neither party, is that the matter of whether ailerons cause a turning effect is an exceedingly complex aeronautical question that depends on many factors, including the aircraft's speed and prevailing wind conditions. But Curtiss is technically accurate on the main point: that ailerons can function completely independently of the rudder as the system for an airplane's lateral control, unlike the patented design of the early *Wright Flyer*.

In any event, the argument is far beyond the capacity of Judge Hazel to adjudicate. Even more important, it is far afield from the real heart of the issues at stake, namely, what kind of exclusive claim should an inventor deserve for his or her seminal idea and how broadly should that claim be construed? Did Selden's half-baked, unrealized brainstorm entitle him to decades' worth of royalties from all car manufacturers? And, if not, what should the Wrights be entitled to for their early-but-incomplete insights about the lateral control of an airplane?

Even before the lawsuit had officially gotten under way, Curtiss wrote the Wrights expressing his hope that the dispute between them could be settled discreetly and amicably. "I suggest," he wrote, "that the matter be taken up privately between us to save if possible annoyance and publicity of lawsuits and trial."

But the Wright brothers are bitter and obstinate. An invention they alone nurtured secretly for years is now advancing at a breakneck pace and, to them, the situation is intolerable. All they can think to do is to try to stop the interlopers.

It is hard to understand the Wrights' motivation in pursuing such a litigious path. Perhaps the root cause of their obstructionist stance is nothing more than an issue of control. As Wilbur laments in a letter shortly before his death, "It is always easier to deal with things than with men." But, whether the Wrights like it or not, the flying machine has burst out of the workshop and into the sky. From now on, it is destined to be adapted and perfected by an impetuous and driven array of inventors around the world.

Grover Loening, close associate of the Wrights, claims in his memoirs that their battle against Curtiss boiled down to nothing more for them than a consuming fight "for revenge and prestige."

In any event, the Wrights were motivated by more than money.

After all, by 1910, the Wrights have become wealthy from the air-plane. They have successfully sold their aircraft to the Army and to professional pilots, and they have received payments from licensed manufacturers around the world. By November 1910, the Wrights' company is incorporated with $1 million in capital from Wall Street investors. The new company even buys the all-important wing-warping patent from the Wrights for an additional $100,000. Yet, even with this windfall, Wilbur, the new company's president, con-tinues battling in the courts.

Curtiss is the Wrights' prime target in their desperate fight to own the skies, but he is by no means the only one. When the color-ful French pilot Louis Paulhan comes to the United States to fly in the 1910 Los Angeles air meet, for instance, the Wrights' attorneys meet him as soon as he steps off the boat in New York. The lawyers put him on official notice that the Wrights are seeking to legally enjoin him from flying in the United States. Not surprisingly, the Wrights' dramatic lawsuit spawns an outpouring of editorial sym-pathy for Paulhan. Newspapers in the United States and abroad express outrage that the country should greet a guest to an interna-tional air meet in such a manner.

Paulhan does fly in the Los Angeles meet on January 10, 1910, but the Wrights' injunction against him puts a stop to his plans to tour the country making demonstration flights before paying audiences. In an arrangement with a promoter, Paulhan's flights had been pub-licized across the country from San Francisco to St. Louis. Paulhan had even been guaranteed an unprecedented fee of $24,000 per month for his performances. In classic French style, he had brought with him no fewer than four airplanes (two Farman biplanes and two Bleriot monoplanes) and an entourage including his wife, two assistant pilots, and five mechanics.

Sometime after the Los Angeles meet, when he receives word that the Wrights have won the injunction against him in U.S. district court, Paulhan clenches his fists and curses the brothers roundly in his native French. Only slightly more demurely, to reporters, he calls the Wrights "birds of prey."

Psychologically, at least, Curtiss benefits from the growing anti-Wright sentiment. Members of both the U.S. Congress and the Justice Department begin to discuss the prospect of putting an end to the "air trust." Fearing the effect on the new industry, the Aero Club of America—still the leading group of aviation aficionados of the day—floats the idea that they might purchase the U.S. rights to the Wright patent and dedicate it to the public domain. But the Wrights' Wall Street backing is predicated largely on the promise of a monopoly on aviation, and the Wrights show little enthusiasm for the plan.

Around this time, a newspaper cartoon appears in a number of newspapers. In it, the Wright brothers stand side by side on the ground sneering and shaking their fists at an airplane above them. The caption reads: "Keep out of my air!" This kind of widespread outrage at the Wrights can't help but offer Curtiss some consolation as he struggles as a prime casualty of their wrath. But it does little to overcome the dire nature of his immediate circumstances.

For Curtiss's company, things could hardly be worse. First of all, dispensing with Herring proves a messy business. The company board is forced to pass a resolution requiring Herring to hand over any patents or patent applications he has. Failing to comply (because he has none)—and with the board threatening to oust him from the company—Herring actually slips out of the meeting and quietly flees Hammondsport with his attorney. Local headlines trumpet Herring's "sensational getaway." And even then Curtiss has not heard the last of him. Astonishingly, Herring has the gall to sue Curtiss. He claims that he was wrongly dismissed and, most laughable, that he, Herring, is the rightful inventor of much of the company's technology. He seeks to recover purported losses incurred while associated with the operations in Hammondsport. Herring will not prevail, but the largely frivolous suit will drag on for years to come.

During this period, as Curtiss associate Lyman Seely puts it later, "Curtiss was hounded by day and by night with lawsuits and injunctions. Anyone with a heart of less than international championship caliber must have broken under the strain. Far from that; Curtiss continued his aviation work along half a dozen lines apparently unruffled."

Early in 1910, the debacle with Herring is a major distraction for Curtiss. But it is the Wrights' injunction that seals his company's fate. When Curtiss is forced to borrow funds on his personal credit to meet his company's payroll, he has to face the prospect of bankruptcy. He knows he can't hope to sustain an airplane company that can neither manufacture nor exhibit airplanes. The one saving grace for Curtiss is that the firm does not own the physical factory and its few machines. Rather, these are held in his own name, and it is the firm that is bankrupted. With these few assets, Curtiss is left to start over—if he can find a way to persist in the face of the Wrights' near monopoly on aviation in the United States.

Explaining his injunction, Judge Hazel has ruled that the claims of the Wrights' patent "should be broadly construed." Furthermore, he writes, the dissimilarities in the control systems between the Curtiss and Wright planes are all but irrelevant because they achieve "the same functional result." As he puts it: "The patent in issue does not belong to the class of patents which requires narrowing to the details of construction."

Judge Hazel also seems swayed by the Wrights' piracy charge: that the AEA stole ideas from the Wrights and imitated their machine. He refers pointedly, for instance, to Tom Selfridge's request for advice on glider construction even though the request had virtually nothing to do with the Wrights' patent or maintaining lateral stability in flight.

From the first, Curtiss has vowed to appeal the injunction. His determination rests both on his desperate desire to keep his business afloat as well as on a matter of principle—not unlike Ford's stance in the Selden case. Nor does the connection to the Selden case go unnoticed at the time. "The Wrights," Harry Genung notes, are acting like "the Seldens of aviation," demanding that "the world should pay them bounty." Meanwhile, Genung explains, despite the fact that Curtiss's airplanes are technologically far advanced of the Wrights' designs, the brothers are "so stubborn in their demands for control, etc., as to make the thing ridiculous from our point of view."

As Curtiss tells the press that winter: "The facts are, the honors bestowed upon the Wrights and the credit which is due them is not because of their patent, but because of their achievement. I hope their patent will be adjudicated and that the machines which operate as their machine does [i.e. wing warpers] will be adjudged infringements, but all authorities agree that our machines operate on a different principle and I do not believe a disinterested patent attorney could be found who would allow that our machine infringes their patent."

Sometime later, Curtiss will reflect further on his actions. "Had we not taken this stand, he writes to a colleague at the end of 1912, "the Wright Company would have been in position to enjoin all manufacturers and the whole industry would have been monopolized."

Early in 1910, seeing few other options and desperate for a resolution, Curtiss tries again to negotiate with the Wrights. According to one account, shortly after the injunction, Curtiss boards a train to Dayton to meet the Wright brothers and their sister Katharine at their home. In person, Curtiss hopes, they can reach some kind of accommodation. But the Wrights, formal and solemn, are in no mood to negotiate. They flatly demand a 20 percent royalty on the retail price of every plane Curtiss sells and on every dollar he earns through exhibitions. Worst of all, they demand that Curtiss pay them royalties on all past money he has earned, arguing that all the airplanes he has built and prizes he has won infringe upon their patent.

Their unyielding position and stony delivery leave Curtiss speechless. He can think of little to say except that he will take the matter under consideration and the meeting ends abruptly. As the story goes, Curtiss wanders the unfamiliar streets of Dayton until late that evening when he decides to telephone his stalwart adviser Judge Monroe Wheeler from an open drugstore. Wheeler corroborates Curtiss's view that he can never accept such terms and hope to survive in the airplane business. He urges Curtiss to come straight home so they can think the situation through.

So Curtiss heads home. But, unlike the mechanical setbacks that he has excelled at overcoming, the legal morass leaves him stymied. He doesn't know what his next step can possibly be. His company is in a financial shambles. He can never afford to compete against the Wrights if he has to pay them $1,000 of every $5,000 airplane he sells. It is an untenable demand and they know it. Meanwhile, he can't even put on exhibitions before paying spectators to raise the money to fight the Wrights in court.

Half a year earlier, Curtiss had proved himself the best airplane designer in the world. Now he faces bankruptcy and the prospect of being run out of aviation entirely.

NINE

FLIGHT OF A HERO

When everything seems to be going against you, remember that the airplane takes off against the wind, not with it.

—HENRY FORD

Even his extreme persistence and optimism cannot alter the dire circumstances Curtiss faces in early 1910. His company is forced to file for bankruptcy. He is forbidden from selling airplanes, pending resolution of his legal battle with the Wrights. He can't even exhibit them.

With the help of his friend and adviser Judge Monroe Wheeler, Curtiss does get the court to allow him to post a $10,000 bond and resume aviation work while appealing Judge Hazel's injunction. The money, in essence, will serve as an advance on royalties due to the Wrights in the event that Curtiss loses his case. Curtiss finds the money to post the bond, but he is forced several times to make payroll out of his own dwindling pocket. Even worse, given his precarious legal situation, he doesn't know where he can turn for a loan, and his company cannot pay all its creditors.

In reviewing his options with Wheeler, Curtiss recognizes an opportunity in the most tantalizing aviation contest of the day. Joseph Pulitzer, the wealthy publisher of the *New York World,* has offered a $10,000 "Hudson-Fulton Prize" to the first aviator to fly from Albany to Manhattan. The newspaperman's prize money could help keep Curtiss's aviation business alive during the lawsuit with the Wrights.

There is a problem, though: almost everyone deems the flight from Albany to New York City to be impossible. Not one airplane pilot has stepped forward to try to meet Pulitzer's challenge. In fact, no one in the country has ever attempted this kind of flight from one city to another.

Curtiss doesn't know if he can do it, but he does know that he desperately needs the money. And, impossible or not, it is one of very few promising options he sees. Even more than the allure of the prize money, he comes to see the Hudson River flight as a kind of redemptive project—a way to somehow surmount his legal woes.

Curtiss first raises the subject one evening at the dinner table with Lena and the Genungs. Harry and Martha Genung, loyal as ever to Curtiss and still living in the back of his house, listen intently to Curtiss's idea as does Lena. And the three unite to try to discourage the outlandish notion. The flight, they argue, is too risky.

Lena, who has always supported Curtiss's hazardous forays into aviation, worries this time. A flight from Albany to Manhattan, she says, tempts fate. Not only does it mean flying some 150 miles over unknown territory; she shudders to think of the perils of piloting an airplane over water. On this latter point, Lena's concern helps sharpen Curtiss's thinking. Hoping to win her over, he says that he will attach flotation devices to his new airplane.

The more Curtiss thinks about this idea, the more he likes it. If successful, he realizes, such a design could make the river—and all bodies of water—relatively safe terrain to fly over.

Before long, Curtiss draws Kleckler into the secret project, and the two begin to work out the technical details of building an airplane that can go the distance—and float on water. Curtiss remains cognizant of the dangers involved in his proposed adventure, but as he and Kleckler tackle the particulars and calculate the aircraft's requirements, he becomes ever more confident.

No sooner has Curtiss set his mind on the Albany–Manhattan flight than he realizes the intricate planning it entails. Compared with even a lengthy aerial demonstration at an exhibition, the flight means facing unknown risks both in the air and below. Curtiss arms himself with maps, weather data, and sketches. He makes several trips along the Hudson by train and boat to study the route. In signature fashion, he becomes a juggernaut of action even as those closest to him remain skeptical.

Over the next few weeks, Curtiss learns as much as he can about the wind patterns along his proposed flight path. By contacting the U.S. Weather Bureau, he determines that the prevailing winds in the Hudson River valley are from the northwest. Based on this information, he decides to make the flight southward from Albany. It feels like the right choice. Plus, he knows from experience that engine trouble is most likely to occur soon after takeoff: the Albany end of the trip will afford far more open space if he needs to make an emergency landing.

Meanwhile, Curtiss and Kleckler try a number of ideas before settling upon the design for the new aircraft's flotation gear. Ultimately, they fit an airtight metal pontoon beneath each wing and, from Baldwin's balloon cloth, they sew five small, inflated air bags onto the undercarriage of the airplane's frame. As Curtiss notes later, the airplane he will dub the *Albany Flier* is the world's first "amphibian plane." It cannot take off from water (Curtiss will solve that problem later) but, as he and Kleckler will soon prove in tests on Lake Keuka, it can handily accomplish a water landing. And, of utmost importance, the *Albany Flier,* boasting the most powerful motor Curtiss and Kleckler have yet produced, is strong and steady in a variety of air conditions.

That April, having posted the bond in the Wright lawsuit, Curtiss is free to head south to test the new airplane in a series of exhibition flights. Lena accompanies him and, at an aviation meet in Memphis, he takes her up in the new machine as a passenger. As the couple flies high above the Memphis fairgrounds, Lena marvels at the world spread out beneath her like a patchwork quilt of greens and browns. It is her first and only flight in an early plane. "I enjoyed it immensely," she tells reporters later. "I wasn't one particle afraid." And, as Curtiss has undoubtedly hoped, the experience quells some of her concern about his plans for the extended Hudson River flight.

Giving them just a few days' notice, Curtiss telephones the Aero Club of New York, designated by Pulitzer as official observers for the flight. He notifies the club officers that he will attempt to fly for the prize on Thursday morning, May 26. He also sends a formal announcement to the *New York World.* "I have made exhaustive experiments with the object of perfecting a machine that would start from and alight on the water," Curtiss writes. The work, he notes, is not yet completed, but he hopes to give his new aircraft "its first practical test by attempting to fly from Albany to New York over the Hudson River, with the hope of obtaining the *World*'s most commendable Hudson-Fulton Prize."

The news sparks headlines and much excitement. The *New York World* launches an immediate publicity campaign for the flight, including a display in the lobby of the Pulitzer Building of models and photographs about the flight and the history of aviation. Not to be outdone, the rival *New York Times* announces a coup: it will charter a special train on the New York Central's Hudson River Line to pace the flight, carrying Mrs. Curtiss and other members of the Curtiss team. Much to the dismay of the staff at the *World*, the train will also carry *New York Times* reporters and cameramen, affording them an exclusive opportunity to keep abreast of the plane every step of the way.

With everything set for the flight, Curtiss sets out for Albany with his small team, including Lena, Henry Kleckler, and a few other Hammondsport hands. Upon their arrival, they check into the Ten Eyck Hotel, whose proprietor, Jacob Ten Eyck, not coincidentally, is the head of the newly formed Aero Club of Albany. Ten Eyck, naturally ecstatic to be privy to the excitement, generously offers every assistance he can as Curtiss and the team ready for the flight.

There is much to be done, not the least of which is to find a suitable spot for takeoff. Back in Hammondsport, Captain Baldwin told Curtiss about a flat plain on Van Rensselaer Island at the southern edge of Albany. With Ten Eyck's directions, the group heads there directly. Low and flat, with an open view of the river, it is ideal. At a meadow beside a cornfield, Curtiss bargains with a local farmer to lease his fallow field for a few days. The farmer suggests a fee of $100, but Curtiss, pleading poverty without having to exaggerate, satisfies the farmer with prompt payment of a crisp five-dollar bill instead.

The group raises a tent at the edge of the meadow and begins to unpack the boxes containing the *Albany Flier*. Leaving Kleckler in charge, Curtiss and Lena set out to find a riverboat to take them down the Hudson for a final inspection. On board, Curtiss plies the crew with questions about the weather and winds along the river, corroborating the information he has already gathered: he will likely encounter the strongest and most dangerous winds past the southern Catskills, by the ominous-sounding Breakneck Ridge and Storm King Mountain.

The close inspection of the route is essential not just to gather information about the wind but also to study the topography. The rules set by Pulitzer allow up to two stops along the route, provided the journey is made within a twenty-four-hour period. There is no thought of a nonstop flight because no airplane at the time could carry the weight of enough fuel to cover such a great distance. As a result, Curtiss needs to reconnoiter suitable landing sites along the route.

After an overnight stay with Lena in Manhattan, Curtiss learns from the officers of the Aero Club that his old friend Augustus Post, now secretary of the organization, will accompany him back to Albany. Post will serve as the club's official observer and is thrilled for the opportunity. The two men turn to marveling at how quickly aviation has progressed. Post begins by recalling that it has been less than two years since he witnessed Curtiss's mile-long *June Bug* flight in Hammondsport. Not too long before that, Curtiss counters, when the AEA moved operations to Hammondsport in 1907, Post was there when the team members wracked their brains about how to get aloft at all.

En route to Albany, Curtiss and Post stop in Poughkeepsie to inspect possible landing sites. Their first prospect—the campus of Vassar College—has too many trees. Far better, they decide, is an area commonly known as Camelot some three miles south of Poughkeepsie with open farmland near the river.

Before they are through, they also visit the large, open grounds of the New York State Hospital for the Insane perched on a hill above Poughkeepsie. There, the superintendent, a Dr. Taylor, shows them around the grounds. As Curtiss later remembers, their guide chuckled "when I told him that I intended stopping there on my way down the river in a flying machine."

"Sure you can land here," Dr. Taylor said, unable to resist a joke. "Most of you flying machine inventors end up here anyway."

May 26 arrives clear and promising, but, true to form, Curtiss will not attempt the flight until the skies are utterly calm. "The weather bureau promised repeatedly, fair weather, with light winds," Curtiss recalls, "but couldn't live up to promises." For the next three days, ready to make an early start, Curtiss wakes at daybreak—normally the time of day with the least wind. On these days, he remembers, the newspapermen and officials, not to mention crowds of curious spectators, "rubbed the sleep out of their eyes" and headed to Van Rensselaer Island. Each day, however, Curtiss recounts, "the wind was there ahead of us and it blew all day long."

On these long, anxious days, Curtiss passes the time intently checking every nut, bolt, and turnbuckle on his machine, tightening them and coating them with shellac to keep them from vibrating loose. As he repeatedly reassures reporters, he has confidence in his airplane, but he will not risk the possibility of strong gusts around

Storm King Mountain and Breakneck Ridge. He will wait for the wind to be just right.

As Curtiss remembers it, "The newspapers of New York City sent a horde of reporters." And, as the days wear on, hard-boiled band that they are, the members of the press become increasingly impatient. As Curtiss notes, "I have always observed that newspapermen, who work at a high tension, cannot endure delay when there is a good piece of news in prospect." Stuck there in the middle of a lonely meadow on a seemingly endless vigil in the boondocks, they badger and taunt Curtiss. One reporter makes a big show of laying odds with the others that Curtiss won't follow through. Others charge that Curtiss is just out for free advertising. About this time, a Poughkeepsie newspaper even gripes about him in an editorial, writing: "Curtiss gives us a pain in the neck. All those who are waiting to see him go down the river are wasting their time."

Nonetheless, the public grows more and more excited as news of the impending flight spreads. Each day, scores of inquisitive onlookers arrive for the chance to inspect the strange flying machine and its pilot up close. Some even set up camp at the edge of the field.

Finally, on Sunday morning, May 29, at the crack of dawn, the air is still. To make sure, Curtiss calls the Poughkeepsie police station and gets just the news he wants: there isn't even enough breeze to flap the flag at the local courthouse. He knows the time has come and prepares to leave Albany immediately.

After a hasty breakfast, Lena heads straight to the train, with Augustus Post, Henry Kleckler, and the others. Except for one *New York Times* reporter and a photographer, both of whom barely manage to rouse themselves in time, they will be the only passengers aboard.

Out at Van Rennselaer Island, Curtiss puts on his flying outfit in the makeshift tent. He will make the flight in a pair of fisherman's rubberized waders that come up to his armpits, a cork life jacket, and a snug-fitting cap. Following Bleriot's example, he also dons a pair of goggles. It is a look that, for years hence, will become de rigueur for pilots. As for the waders, Curtiss later explains that they are not intended so much for the prospect of a water landing as for the warmth they will provide. After all, despite the warm day, he will be flying in the open, hundreds of feet in the air at a speed of roughly fifty miles per hour.

Curtiss is all business now, but he feels a tinge of regret that both Kleckler and Lena, having gone directly to the train, aren't present to lend him moral support. He makes do with the competent help of Clarence White and Elmer Robinson, two younger mechanics who have accompanied him on the adventure. By this point, after all, not much remains to be done; Curtiss has checked and rechecked every inch of the airplane countless times. With no fanfare, he takes his seat in the front of the airplane.

Even with all the delays and false starts, not to mention the early hour on a Sunday morning, at least a hundred spectators are on hand to see him off. Unbeknownst to him, it is but a dim preview of the public interest that lies ahead.

As Curtiss recalls, the extended delays had "gotten somewhat on my nerves." But with the morning calm and bright, he resolved "it was now or never." From his perch on the makeshift runway, he notes the direction of the smoke from factory stacks to judge wind direction and readies for takeoff. According to Jacob Ten Eyck, serving as the official Aero Club starter, Curtiss's wheels leave the ground at exactly 7:02 A.M. As planned, Ten Eyck signals his assistant to wave a white flag from the top of a nearby warehouse visible from the train. At the sight of the flag, the engineer of the train blows its whistle and heads off in a synchronized start down the tracks along the east side of the Hudson.

By wire, the whole Hudson River valley quickly learns of Curtiss's successful takeoff. In Poughkeepsie, according to one account, the bell in the city hall steeple chimes with the news, while in Manhattan people huddle around bulletin boards at the offices of the *World* and the *Times* and in front of smoke shop windows to see phoned-in reports of Curtiss's progress. Anticipating a glimpse of the spectacle, people also begin gathering along the waterfront parks of the city and on the roofs of apartment buildings that offer a view of the Hudson.

Curtiss rises to an altitude of 700 feet and flies straight down the middle of the river. With the Hudson spread out below him like a wide, glimmering road, he notices with fascination that from above he can see through the clear water to deep beneath the river's surface.

"I felt an immense sense of relief," Curtiss would write later, to be finally airborne on such a beautiful, cloudless day. "The motor sounded like music."

It is clear sailing for the first leg of the journey. The machine handles perfectly. When the train first comes into view alongside the river, Curtiss veers toward it to fly alongside. He can see Lena leaning out the window, waving her handkerchief and later a large American flag. Henry Kleckler, too, pops in and out of the train window, nervously flapping his cap. With both train and airplane traveling at roughly fifty miles an hour, Curtiss remembers: "It was like a real race and I enjoyed the contest more than anything else during the flight." The train and the airplane weave together and apart along the voyage. Sometimes, as the tracks move away from the bank or the train slows around curves, Curtiss flies far out ahead only to find that the train is back by his side on straightaway stretches of track along the river.

With little instrumentation, Curtiss has no way to determine his speed other than the strength of the wind against his face. Because he has no altimeter, he can similarly only guess at his altitude. And the deafening drone of the engine behind his head shuts out all other sound.

Nonetheless, he feels in complete control of the airplane and intensely alert to the tiniest details around him on the crystalline day. Below him, Curtiss sees groups of people staring from the riverbanks and boaters waving; the captain of a river tugboat toots its horn; although Curtiss can't hear it, he sees the blast of white steam rise eerily silent into the air below him.

Sooner than expected, Curtiss sees the distant outline of the Poughkeepsie bridge spanning the Hudson eighty-seven miles from Albany, roughly marking the halfway point of his journey. He must land to fill his tank with gasoline. Peering out from his open-air perch, Curtiss soon spots Camelot's open grassland and bounces to a landing on its bumpy field, where he had arranged to have gas and oil waiting for him. But despite the best laid plans, there is no gas or oil to be found.

It is 8:26 A.M. on a Sunday. Perhaps, Curtiss guesses, the local garage proprietor is at church. If so, however, he would seem to be in the minority. According to a newspaper account, church attendance in Poughkeepsie that day falls off so sharply that the Reverend William H. Hubbard of Mill Hill Baptist Church chastises his flock—and Curtiss—for desecrating the Sabbath.

Fortunately, the field in which Curtiss has landed is no more than two hundred feet from the road; already automobiles are pulling over by the dozen, and a crowd of almost a hundred excited spectators congregates around the plane. Curtiss explains his predicament and asks if anyone can spare him some gas and oil. At least a dozen offer their assistance. He gratefully accepts some eight gallons of gas and a gallon of oil from two New Jersey motorists who help fill his tank from spare cans in their touring cars.

By this time, the tracking train has long since pulled off on a siding near Camelot, and its passengers come jogging excitedly across the meadow. Henry Kleckler, like a doctor rushing to the scene of an emergency, is in the lead. He inspects the *Albany Flier* with meticulous care, checking the engine and each of the aircraft's wires and struts. Lena rushes to her intrepid husband and affectionately takes his measure. Post adds a handshake and a pat on the back, and the news photographer snaps a picture of Curtiss standing beside the airplane. And, within the hour, Curtiss is back in the air, his ears still ringing from the roar of the motor.

As he recounts: "Out over the trees to the river I set my course, and when I was about midstream, turned south. At the start I climbed high above the river, and then dropped down close to the water. I wanted to feel out the air currents, believing that I would be more likely to find steady air conditions near the water. I was mistaken in this, however."

As Curtiss drops down close to the water, a gust of air tips his wing dangerously high and he almost touches the water. "I thought for an instant that my trip was about to end, and made a quick mental calculation as to the length of time it would take a boat to reach me after I should drop into the water," Curtiss remembers.

Yet even worse trouble lies ahead.

Twenty miles south of Poughkeepsie, the river carves a steep fifteen-mile-long gorge in the so-called Hudson Highlands near Storm King Mountain and Breakneck Ridge. The spot funnels treacherous wind currents above the river. Aware of the danger from his research and reconnaissance, Curtiss tries to climb above it, rising to an altitude of roughly 2,000 feet. But it is not high enough. Just past Storm King Mountain, as Lena watches, increasingly frantic and helpless, from the train, a cross current tilts the *Albany Flier* sideways, and the plane drops more than a hundred feet within seconds. Momentarily losing control, Curtiss is nearly thrown from the airplane. "It was the worst plunge I ever got in an aeroplane," Curtiss says later. "My heart was in my mouth. I thought it was all over."

As the wind steadies, Curtiss is shaken but he regains control of his aircraft. Ahead, he can just make out the northern tip of Manhattan and the outline of the fifty-story-high Metropolitan Tower—the world's tallest building—above the line of the horizon.

Just as Curtiss begins to feel elated that he is so near the end of the trip, he notices that his oil gauge reads perilously near empty. Like all of Curtiss's airplanes to date, the *Albany Flier* requires the pilot to lubricate the engine through a manual control. While in flight, Curtiss must pull the lever on a hand-operated oil pump roughly every ten minutes to assure the smooth running of the engine. With his recent ordeal at Breakneck Ridge, Curtiss's first thought is that he must have inadvertently "been too enthusiastic" with the oil lever. In fact, although he won't discover it until later, the airplane has been seriously leaking oil for some time. In any event, with the prospect that his engine could freeze up at any time, Curtis knows he must quickly land to replenish his oil.

Nervously winging east at the northernmost tip of Manhattan

where the Harlem River curves around at the Harlem Gorge to meet the Hudson, Curtiss looks for a little meadow at Inwood—one of many such spots he has chosen as possible landing sites. There is no time to lose. Spotting nothing more suitable, he makes an emergency landing on a sloping lawn that rises a hundred feet above the Hudson. Safely on the ground, he breathes a sigh of relief and realizes that he is inside the city limits. It is 10:35 A.M. In just over two and a half hours of flying time, he has covered 137 miles, averaging nearly 55 mph.

Curtiss has landed on the grounds of the estate of the late financier and leather merchant William B. Isham, now inhabited by Isham's daughter and her husband, M. P. Collins. They jump up from reading the Sunday newspaper when they hear the roar of the approaching motor. They have just been reading about the proposed flight and are stunned to see Curtiss's airplane bouncing up their sloping front lawn. Collins rushes to greet the unexpected visitor. Holding out his hand to Curtiss, he says he is delighted to be the first to welcome him to the city and congratulate him on his successful journey. "I am also glad," he says, "that you picked our yard as the place to land."

At the Isham estate, Curtiss telephones the *New York World* with the news that he has landed within the city limits. He will, he says, continue on to his planned landing site at Governors Island as soon as possible.

Having technically fulfilled the contest's requirements—and with only one stop instead of the two allowed by the rules—another aviator might have pronounced the flight complete. But not Curtiss. He says later that he thought of all the spectators in the city counting on his arrival. Doubtless, he thought also about the thrill of being the subject of their adulation. Regardless of his motive, as one magazine writer noted, his decision to fly on over Manhattan was "a magnificent sportsmanlike thing that won him the unbounded admiration of all New York." Having refilled his oil tank, he lifts off at eighteen minutes before noon, heading for Governors Island and glory.

After a dangerous and tricky takeoff down the sloping cliff over the river, Curtiss once again rises over the open Hudson, this time with the shimmering Manhattan skyline beckoning him onward through the clear midday sky. As he approaches the city, he is overwhelmed by the reception. Crowds are everywhere: on rooftops, in trees, and packed many deep along the riverbanks. Passengers on ferryboats and ocean liners crane to railings and wave wildly in the air to him. And people on scores of crafts large and small dotting the Hudson cheer him on as well.

"New York can turn out a million people probably quicker than any other place on earth, and it certainly looked as though half the population had flocked to Riverside Drive or out onto the rooftops of the thousands of apartment houses that stretch for miles along the river," Curtiss recalls. As he says later, he had never experienced anything so dramatic and inspiring.

In no time, the Statue of Liberty—Curtiss's sought-after landmark of the finish line—stands close before him. Turning westward, he remembers, he triumphantly "circled the Lady with the torch" and headed as planned for the parade grounds at nearby Governors Island.

It is just past noon and, after a perfect landing, Curtiss emerges from his airplane to cheers from scores of enthusiastic U.S. Army personnel at the small base there. It has been, to say the least, a full morning, including an unprecedented two hours, fifty-one minutes of flying time.

Despite their own timely arrival in Manhattan, the passengers on the *New York Times* train fall behind while trying to make their way to Governors Island from Grand Central Station. Nonetheless, a representative from the *World*—already on hand—rushes up to congratulate Curtiss and make arrangements to bring him back to the city to receive his $10,000 award.

Making his way toward the ferry arriving from Manhattan, Curtiss is reunited with Lena as she rushes ahead of all the others. Perhaps because she was always nervous about the trip, or because she watched him the whole way, including the rocky turbulence at Breakneck Ridge, Curtiss notes that her embrace is a bit tighter than he expects. Nor does she seem shy hugging him with some abandon in front of all the soldiers. Lena is so happy Curtiss completed the trip safely, in fact, she is only too eager to reenact the embrace, as she will remember, "ever so many times" at the request of the news photographers.

After a formal luncheon at the Hotel Astor, Curtiss is escorted to the Pulitzer Building, where the *New York World* holds a press conference open to all journalists—even those from the train-hogging *New York Times*.

Naturally, the New York press crowns Curtiss "King of the Air." The *New York Evening Mail* rhapsodizes about "the courage of the man" who made the flight even though he knew all along that "a broken bolt or some little thing gone wrong might dash him to death."

The fact is, though, the reporters are not sure what to make of Curtiss. He is, of course, an intrepid and accomplished aviator. But he is so reserved and unassuming the reporters have to work hard to make him seem like the glamorous hero their readers must surely expect.

Curtiss "speaks quietly" and "is not at all forward," the reporter for the *New York World* writes. But, he adds, "there lurks within him the element of enthusiasm that goes to make up great adventurers, and it speaks out from his eyes, which are the most expressive part of his face."

The gala event comes two days later: a formal, gentlemen-only dinner in the Hotel Astor's main banquet hall amid chandeliers, gilded sconces, and potted palms. The black-tie affair, hosted by Pulitzer and the *New York World,* is presided over by New York's mayor, William Gaynor, and the elite guest list of roughly one hundred includes New York financiers, military brass, and wealthy aviation enthusiasts.

Although Curtiss is still no public speaker, he is at least getting more proficient at ducking speeches. He tells the assembled diners that he "planned everything about the flight in advance except the possibility of making a speech." Still, he does have something to present to Gaynor. The mayor of Albany, James B. McEwan, gave Curtiss a letter to carry aloft and deliver to his counterpart in Manhattan. Although Curtiss already delivered it in person at city hall, he and Gaynor reprise the delivery for the assembled dignitaries. The letter offers little more than greetings and platitudes, but everyone present recognizes its symbolic significance: it is the first airmail letter ever delivered in the United States.

Gaynor, a gregarious politician, gladly makes up for Curtiss' chronic bashfulness with several speeches of his own and by reading scores of telegrams of congratulations that have poured in from all over the world. Among them, Gaynor reads a statement from

President William H. Taft: "I am intensely interested in what Mr. Curtiss has done," Taft writes. "It seems that the wonders of aviation will never cease. . . . His flight will live long in our memories as having been the greatest."

As an added bonus, Curtiss learns in the ensuing days that, by making the continuous, eighty-seven-mile flight from Albany to Poughkeepsie, he will receive the *Scientific American* Trophy for the third consecutive time. This time, the magazine decides to retire the trophy and give it to Curtiss to keep in perpetuity. Presenting the hefty trophy to Curtiss several months later, Charles Munn, *Scientific American*'s publisher, says that three names will forever be associated with New York's famous river: Hudson, the explorer who discovered it; Robert Fulton, the steamship inventor who revolutionized river navigation; and finally, Glenn H. Curtiss for his epochal flight.

And epochal it was.

Just as Bleriot's channel crossing kindled the imagination of people throughout Europe to the promise of the airplane, Curtiss's flight from Albany to New York City breaks a formidable psychological barrier for aviation in America.

Not only that, Curtiss has flown a total of 152 miles. His successful airborne voyage from Albany down the Hudson River valley and ultimately around the Statue of Liberty will go down in aviation history as one of the handful of flights to change the world. That Sunday, and not just for the hundreds of thousands of witnesses but for many others as well who read or heard of his accomplishment, the airplane, once a novelty, suddenly and all at once presents itself as a useful and practical technology. In one dramatic journey, Curtiss forges a path for the development of airmail, modern air travel, and the terrible prospect of air power in war.

Perhaps most notably, the import of the flight is recognized at once. "Man has conquered the air," the *New York Times* announces. "Accomplished with wonderful courage and skill," Curtiss's flight, "seems to make human flying more a reality than hitherto it has been."

Thanks to Curtiss's feat, the *New York Times* coverage continues, "the development of the airship practically and commercially and the growth of its usefulness as a carrier are only matters of time." In all, the *Times* devotes six full pages of text and photos to Curtiss's Hudson flight. It is the most space the newspaper of record has ever devoted to a single news event.

TEN

NEW BEGINNINGS

*It would be indeed little less than miraculous if the
way to success were not taught by failures at first, here
as everywhere.*

—SAMUEL PIERPONT LANGLEY, 1892

With the Albany–Manhattan flight,
Curtiss becomes a bona fide American hero. For many years to
come, however, the ongoing saga of the Wrights' bitter lawsuit
against him sets all Curtiss's accomplishments against a charged,
highly partisan backdrop. The Wrights loathe his success. They
consider their discovery of wing warping to be the singular development that makes the airplane possible. Not only does Curtiss
rebuke their exclusive claim on this seminal contribution but, from
their point of view, he flaunts his piracy, garnering accolades and
worldwide acclaim.

As a result, the Wrights want not just to fight Curtiss, but to punish and destroy him. It becomes an obsession. People close to the

Wrights note with some alarm how the irrepressible Curtiss becomes a chronic focus for their bitterness. As the Wrights' colleague Grover Loening puts it later, Orville "openly despised" Curtiss and nurtured a "vicious hatred and rivalry" toward him as the lawsuit dragged on.

Before long, in fact, litigation comes to dominate the Wrights' lives as the brothers spend increasingly long periods giving testimony in court and less time updating and improving their airplanes. Their official statements become more shrill. "We shall hold all persons using or engaging others to use infringing machines strictly accountable for all damages and profits accruing there from," their company literature warns. They even threaten to sue the U.S. military on the suspicion that Navy officials are considering the purchase of an "infringing" airplane other than theirs.

Feeding the Wrights' fury, a federal appeals court overturns Judge Hazel's injunction against Curtiss in June 1910, just weeks after the Albany–Manhattan flight. The three-judge panel rules that Curtiss's alleged "infringement was not so clearly established as to justify a preliminary injunction." Plus, the judges note, a number of affidavits in Curtiss's favor have been presented after Hazel's decision and deserve a fair hearing.

Curtiss knows that the legal sparring over an injunction is just a prelude. Now he faces an uphill and sure-to-be grueling legal battle on the merits of the patent infringement case—with Judge Hazel once again presiding at the bench. But, for the moment at least, he is legally free to continue in the aviation business.

With the injunction lifted, and riding high on the accolades—and prize money—from the Hudson River flight, Curtiss seizes the chance to start again. He forms a new company from the tattered remains of the debacle with his former partner Herring.

Curtiss has, by this point, learned something of the rough and volatile world of business. He makes far sounder choices this time in establishing two new firms—the Curtiss Aeroplane Company and the Curtiss Motor Company. In each case, the company's stock is tightly held by a small, trusted group including Lena Curtiss, Monroe Wheeler, Harry Genung and Henry Kleckler. As he sets out anew, Curtiss will need every bit of the business acumen he has acquired to keep the company afloat while *Wright v. Curtiss* proceeds through years of affidavits, expert testimony, and mounting legal fees.

Underscoring the sense that things are on a more secure and optimistic footing, Curtiss becomes a father in June 1911. His son—Lena insists on naming him Glenn Jr.—is born nine years after the traumatic death of their firstborn infant. Lena is thrilled to learn her new son is healthy.

During this period, much of the income of the Curtiss Aeroplane Company derives from exhibitions. After the Albany–Manhattan flight, it seems, the entire country wants to see Curtiss fly. Nonetheless, given Curtiss's new paternal role, Lena urges him to let others do the exhibition flying. Training a strong new crop of young pilots, Curtiss agrees to curtail his own flying, at least over land, so as to focus on the inventing he loves best.

The next few years are Curtiss's most productive, despite the drain on his time and money from the Wrights' lawsuit. His prodigious stream of inventions surely rivals any in the annals of aviation. Curtiss invents the hydro-aeroplane—now called the seaplane—including a pontoon design, still in use today, that handily allows the aircraft to break away from the surface tension of the water upon takeoff. He masterminds the first launch of an airplane from a ship, presaging the modern aircraft carrier. He is the first to design retractable landing gear on an ingenious aircraft he calls *Triad* that can take off or alight on land or water. He perfects a design for a "flying boat" in which the airplane's fuselage itself doubles as a hull for landing on water. He works with Elmer Sperry Jr. to help develop a gyroscopic automatic stabilizer for airplanes. All the while, he designs and builds ever stronger engines. And, perhaps most important to the future of his company, he lays the foundation for a new airplane, the *Curtiss JN,* that will come to be known as the "Jenny"—one of the most popular and successful airplanes in the history of aviation.

Throughout, Curtiss retains his characteristic openness and optimism. Theodore "Spuds" Ellyson, who works closely with him over these years, later recalls his extraordinary focus and determination. In his initial design for the hydro-aeroplane, for instance, Curtiss and his team experiment with some *fifty* different pontoon designs before they can get one to work satisfactorily. "Those of us who did not know Mr. Curtiss well," Ellyson says, "wondered that he did not give up in despair."

Those who do know Curtiss, however, know he is not the type to despair, no matter what the circumstances. He thrives on problem solving and experimentation. And he is good at it. Navy Captain Washington Irving Chambers, who worked with both the Wrights and with Curtiss notes that Curtiss was "always ready to make experiments and was as progressive as the Wrights were conservative." Chambers describes Curtiss as "a mechanical wizard" who would "speak little as he attacked a problem with logic and at-hand materials, cutting and trying until it worked."

"Experimenting is never work—it is plain fun," Curtiss once

explained to a reporter. He could work eighteen hours a day when occasion required it without feeling any special fatigue, Curtiss said, because he always spent at least half of his time experimenting with new ideas. He called the process "rejuvenating," adding that "the man who works himself down can work himself back up if he will develop a turn for experimenting."

Coupled with Curtiss's love of experimenting is the same straightforward, competitive verve that led him to championships as a bicycle and motorcycle racer years earlier. Curtiss retains that driven, restless energy throughout his work in the aviation field. Perhaps most noteworthy, though, especially in light of his treatment at the hands of the Wright brothers, Curtiss always shares the fruits of his inventive imagination freely with others. Over the course of the years in litigation with the Wrights, Curtiss, advised by his attorneys, does register some aviation patents—such as a patent on the flying boat. But it is a small handful compared with the estimated five hundred aeronautical inventions he is ultimately credited with. And, even when he does patent his inventions, he never once uses his patents as a weapon to control the technology's development or shut out competitors.

Nor does Curtiss, unlike the Wrights and Henry Ford, ever become bitter from the chronic legal machinations he faces. When others copy his machine—even openly selling "Curtiss-type" aircraft—Curtiss pays practically no attention. As he tells his colleague Lyman Seely, all the lawsuits convince him that the best approach is "to forget about patents and look for the business." The goal, he says, ought to be simply to keep building better airplanes than anyone else.

The strategy earns Curtiss even more recognition and success. He wins every top honor in the field for his accomplishments including, in 1911, the Aero Club's newly established Collier Trophy—the highest award for achievement in aviation—and, in 1913, the Langley Medal, an award the Smithsonian had formerly bestowed only upon the Wright brothers.

Ironically, even in these years while the lawsuit moves glacially forward, the Wright Company will appropriate some of the best of Curtiss's inventions. Purportedly as an aid in their lawsuit, Wilbur and Orville quietly procure a Curtiss plane, study its construction, and fly it repeatedly in Dayton. Before the lawsuit is over, in fact, Wright Company planes will abandon the wing-warping system in favor of ailerons, the U.S. patent for which is finally granted to Curtiss and the AEA in 1914, after being held up for years at the patent office by the Wrights' lawyers. By that year, the Wright Company even starts selling "flying boats" that look suspiciously similar to Curtiss's patented design.

But Orville will not stop the lawsuit against Curtiss over the antiquated wing-warping technique—especially after Wilbur dies in 1912. And, with strong legal representation, Orville prevails in court. On February 1913, Judge Hazel rules against Curtiss and this time, in January 1914, Curtiss loses his appeal in the case as—crushingly—the federal appeals court upholds Hazel's broad interpretation of the Wrights' wing-warping patent.

In retrospect, the verdict says more about the strength of the Wrights' legal team and the woeful failure of the U.S. patent system than it does about the rightful progenitor of aeronautical technology. Here again the Wrights' secrecy played a role. Had the Wrights, back in 1906, been willing to demonstrate their airplane to the patent examiner, they might have garnered a patent on their particular airplane design itself, as so many airplane designers would subsequently do. But, of course, the Wrights were unwilling to show anyone their airplane, and the patent office was unwilling to hand out patents on flying machines without proof that they could work.

Of necessity, in the years following their 1903 success at Kitty Hawk, the Wrights' clever legal team gambled that the brothers' patent might be even broader and more invincible if it focused on the principle of lateral control embodied in the early *Wright Flyer*.

The improbable strategy paid off.

Now, the 1914 federal appeals court verdict against Curtiss places him in an all-too-familiar position. Once again, his options seem impossibly few. And his future in the airplane business stands at the mercy of his bitter adversary Orville Wright. Curtiss doesn't want to be bankrupted a second time. Nor does he want to move his company to a foreign country—another of his possible options. But he has little more recourse through the U.S. courts.

Normally, at the end of such a titanic legal battle, the winning company would negotiate terms to its own advantage. But, to Orville, it seems, hurting Curtiss is more important than helping his own firm. He sticks to his demand for 20 percent royalties on all airplanes Curtiss has ever sold or exhibited, knowing full well that the demand will likely bankrupt him once again.

Orville's vindictive strategy is well recognized at the time. "It is difficult to believe that Mr. Wright is actuated by other than personal animosity," notes an editorial in the *Boston Transcript* in 1914, for instance. "He must see that his course will not pay in a financial sense, for under his terms he can secure no licenses, as he could easily do under a more liberal policy."

Not knowing where else to turn, Curtiss decides in the spring of 1914 to take up Henry Ford's offer for assistance and Ford is eager to help. "My entire legal staff is at your disposal," he tells Curtiss.

"Patents," Ford says, "should be used to protect the inventor, not to hold back progress."

From the start, Ford's brilliant attorney Benton Crisp begins to reap rewards for Curtiss. Initially, in the summer of 1914, he devises a legal stalling tactic. Noting that the Wrights' wing-warping patent specifies that the wings must be twisted simultaneously, he helps Curtiss devise an "interlock" for his airplanes' ailerons making them operate independent of one another on each of the airplane's wings. These so-called nonsimultaneous ailerons could be seen as an advance that would not be covered by the Wright patent. As expected, Orville is not satisfied with the change. But the tactic profitably moves Curtiss back to court for yet another round.

In the meantime, characteristically, Curtiss looks for another show stopper like Rheims, or the Hudson River flight to somehow transcend his legal woes. The idea for a flight across the Atlantic Ocean fits the bill nicely. So does the notion, first broached by Lincoln Beachey according to one account, to see whether a new trial of Langley's aerodrome—the ill-fated dragonfly—might somehow dent Orville Wright's expansive legal claims over the airplane. After all, Samuel P. Langley's aircraft predated the *Wright Flyer;* the courts' decisions in favor of the Wrights hinge largely on the fact that their patent's pioneering status merits an expansive interpretation.

At the end of May 1914, Hammondsport is buzzing. Work on the proposed Atlantic-conquering *America* is already under way. Zahm, Manly, and Walcott are in town and, after nearly two months of work, the first phase of the aerodrome reconstruction is nearing completion.

As the final repairs to the aerodrome are made, workers from the

Curtiss shop carry its pieces to a hangar near the shore of Lake Keuka. Late into the evening, they work to assemble Langley's restored aircraft for launching.

The next day, May 28, 1914, Curtiss and the restoration team are back at work by sunrise.

It is a clear morning with a mild, shifting breeze. Ready for action, Curtiss surveys the scene from a familiar spot at the southern tip of the long and narrow Lake Keuka. The verdant, tree-lined hills rise from the lake's distant banks in dark silhouettes against the rosy morning light and the day's first rays of sunlight glint off the water.

From this exact location just seven winters ago, Curtiss first felt the thrill of aeronautical success when Casey Baldwin made his first brief flight in the AEA's *Red Wing*. From this very spot, over the ensuing years, Curtiss has tested dozens of other prototype aircraft, including the *Albany Flier*, and his early hydro-aeroplanes.

Along the way, as Augustus Post has put it, Curtiss has "worked his way up from the making of bicycles to the making of history." Through his own creativity and perseverance, he designed and built a motorcycle to carry him faster than anyone had ever traveled over land. As a member of Bell's team, he crafted an airplane that flew in public before the Wright brothers did—a plane that incorporated more lasting aeronautical features than any of the Wrights' early designs. In France, he tested his skill as an inventor and pilot against the greatest aviators in Europe and beat them all. He mastered the problem of taking off and landing from the water. And, with his daring flight above the Hudson River, he helped open the eyes of the world to the practical potential of air travel.

Altogether, it is a stunning string of successes for a man just past his thirty-sixth birthday.

But this morning's flight is different. For once, Curtiss and his team won't be forging new technological ground but looking backward, promoting an appreciation of the fact that sometimes a seminal technological "failure" can be as important as the successes that may follow it. Already, the reconstruction effort has paid homage to Langley's brilliant, creative work. Now, perhaps, a successful flight in Langley's rebuilt aerodrome will restore his reputation and forever alter the way the story of the airplane is told.

Just a decade earlier, the aerodrome had brought Langley nothing but ridicule and the world scoffed at the possibility of heavier-than-air flight. Now airplanes have become commonplace. They have transformed the world. And yet the courts continue to wrangle—unproductively—over who should get the rightful credit for the airplane's invention.

Given the charged backdrop of the Wrights' protracted lawsuit against him, Curtiss surely knows that the reconstruction of Langley's aerodrome somehow represents a fight for his own place in history as well. By resurrecting the Langley, Curtiss champions his view that technological breakthroughs emerge not in isolation but by building upon the work of those who have come before. As Zahm will express it in his official monograph for the Smithsonian on the aerodrome reconstruction: "The aeroplane as it stands today is the creation not of any one man, but rather of three generations of men."

Curtiss squats by the lakeshore. There are no whitecaps on the water, but, with a practiced eye, he gauges the wind gusts by studying the movement of the shallow streaks of waves across the lake's surface. Frowning, Curtiss opts, as usual, to wait until the wind dies down to his satisfaction.

Finally, at around 7 A.M., a dozen men, in rolled-up shirtsleeves

and knee-high rubber boots, lift the gangly aerodrome and wade
into the water to set it down carefully on its pontoons. Improbably
enough, alighting on the surface of Lake Keuka, Langley's ill-fated
dragonfly is reborn. Now the craft will get an exceedingly rare
opportunity: a second chance to make history.

From the ankle-deep water, Curtiss climbs aboard the craft to sit,
as Manly had once done, on the wooden board in the little fabric-
sided booth under the forward part of the frame that serves as the
cockpit.

By now, many eager spectators, reporters, and photographers
have gathered along the shore, including seven whom Zahm and
Walcott designate as official witnesses. Unlike many of Curtiss's
demonstration flights, a quiet, almost reverential mood descends
upon the crowd. The strange circumstances of the flight and the
antiquated look of Langley's contraption conspire to make everyone not just excited for the outcome but somehow mindful of the
sweep of history.

Into the morning's hush, the original motor that Manly built so
long ago now coughs and roars into life and the aerodrome's propellers start to spin. The four-winged craft, headed across the wind,
starts to skid over the little waves. Then, with its huge rudder acting
like a weather vane, Langley's aerodrome, of its own accord, veers
straight into the wind under full power. In what Zahm describes as
a slow and "stately" takeoff, it rises on an even keel, a few feet above
the water, and heads airborne toward the distant shore.

Feeling disproportionate drag on the aerodrome's left wings, Curtiss cuts the aerodrome's engine and lets the aircraft alight softly
on the water after flying for just 150 feet. As a motorboat tows Cur-

tiss and the aerodrome back to the shore, he explains that he
landed because, unfamiliar with the controls, he feared the craft
might list or even keel over. Despite the flight's brevity, though,
there is no doubt about the outcome: Langley's aerodrome has
finally flown under its own power. Without delay, the reporters on
hand scramble to town to wire their stories from the back of Jim
Smellie's pharmacy.

The Curtiss-Smithsonian team's initial feat with an airplane
approximating Langley's original, firmly establishes the viability of
the aerodrome and makes headlines around the world. The next
day, a front-page article in the *New York Times* declares, for
instance, "'Langley's Folly' flew over Lake Keuka today—approxi-
mately eleven years since it caused the country to laugh at its inventor when on its trial flight it fell into the Potomac. " As the *Times*
reporter puts it: "It was one of the greatest days for aviation that the
town has known."

Not surprisingly, the *Hammondsport Herald* goes even further,
suggesting that a review of the aviation field "reveals the fact that
practically every machine in existence traces its ancestry quite
directly back to the genius of Dr. Langley."

News of the flight of the rebuilt aerodrome particularly pleases
the aging Alexander Graham Bell. Bell had long admired his colleague Langley, whose research had inspired him and so many others to enter the emerging aeronautical field. "Congratulations on
your successful vindication of Langley's drome," Bell wires Curtiss
upon hearing the news. "This is really the crowning achievement of
your career, at least so far."

That first day, no more tests can be made on the aerodrome
because, when the team returns to the lake after an excited breakfast, they discover that a bearing on one of the propeller shafts has

given way, requiring lengthy repairs. Over the next few days, however, Curtiss lifts the machine off the water for several more short flights, accommodating photographers and proving that his initial result can be successfully replicated. On June 2, Smithsonian Secretary Walcott, beaming, officially pronounces that the group has proven the viability of the original Langley aerodrome.

Before returning to Washington, Walcott orders the next phase of the work to begin with an installation of a Curtiss engine and propeller to further test the aerodynamics of Langley's tandem-wing design.

The second phase of the experiments—to gather data on the potential of the tandem-wing design—will be placed on hold for the moment while the high-profile *America* moves to center stage. But ultimately, by the summer's end, the aerodrome, with a new Curtiss engine and several other modifications, will remain airborne for nearly a half hour, flying in one test for a full ten miles into a stiff wind.

What did Curtiss and the Smithsonian team prove in the Langley reconstruction? The question has been asked and debated heatedly since the experiment was conducted in the spring of 1914 because, for literally decades to come, Orville carries on a campaign to discredit the group's work.

Working with his British friend Griffith Brewer, Orville documents many changes that the reconstruction team made in Hammondsport. So many, in fact, that an exasperated Zahm retorts that he wonders why they don't complain that the aircraft's color changed from white to buff when its fabric was replaced.

Many of the changes Orville gripes about are incorporated only in the later, second phase of the work when the group makes no

secret of the fact that it is introducing modifications to test the aerodynamics of Langley's favored tandem-wing design. The details of
Orville's charges, after all, are gleaned in only in this second phase
when, in two separate trips, Brewer and Orville's older brother
Lorin are dispatched to spy on the reconstruction effort.

Nonetheless, several of Orville's charges are significant. Most
troubling, perhaps, is the fact that, from the first, the Curtiss-
Smithsonian team considerably strengthened the bracing of the
wings—a change that alone may well have forestalled a repeat of
Langley's debacle. They justified the additional bracing, reason-
ably enough, by the fact that the aerodrome had to be able to support the pontoons. But even Zahm ultimately acknowledged that
the 1914 test cannot be seen to definitively prove the structural
soundness of the original Langley aerodrome, only the viability of
its aerodynamics and the sufficiency of its engine.

Equally noteworthy, the reconstruction team omitted the sharp-
edged front extensions to the wings that Langley added to his final
aerodrome model. They opted instead to leave a more rounded—
and aerodynamically sound—leading edge to the wing. It is debat-
able whether Langley's extensions to the wings' forward ribs were
integral to his design. But there is little doubt that the reconstruction team realized their chances of success would be greater if they
ignored this misguided feature. The omission can certainly be seen
to diminish the team's claim to scrupulous scientific accuracy. But
arguably, at least, it should have done little to dent their credibility.
Even with this omission, as they all contended, the reconstructed
aerodrome was substantially as Langley had designed it.

Nonetheless, Orville's charges of alterations to the aircraft were
delivered by Brewer along with sensational allegations of fraud in a
paper presented in 1921 to the Royal Aeronautical Society in Lon-

don. In the view of Brewer and Orville, the reconstruction team knowingly misrepresented their work and perpetrated a fraud on the public in order to steal the Wright brothers' reputation as the world's first to fly.

As Zahm strenuously counters, "To impugn the motives of the Smithsonian men associated with the work of retesting the Langley aeroplane in 1914 is a discourtesy and injustice that well might be discountenanced by an impartial society."

But history, it seems, is not written by an impartial society but, most often, by those who crow the loudest into posterity. Accordingly, Brewer's trumped-up charges largely stick for many decades to come. It does not seem to matter that all the principal team members involved in the reconstruction address and refute most of Orville's and Brewer's claims. Perhaps Zahm answers most forcefully. Challenging Brewer's charge of a conspiracy, Zahm declares that the experiments to reconstruct the Langley aerodrome "were no more initiated for the purpose of patent litigation than were Langley's original experiments." Charges of any kind of a conspiracy, Zahm said, were "the irresponsible gossip of a partisan who could easily have ascertained the truth."

Nor does it seem to make a difference that Curtiss and Manly both separately express to Orville their willingness to undertake yet another test of Langley's aircraft under any auspices he might suggest. Orville chooses to decline the offers.

Looking back, the Curtiss-Smithsonian group can perhaps be faulted for an excess of zeal—as well as an indisputable conflict of interest in the aerodrome affair. All the principal team members certainly wanted to see Langley's rebuilt airplane succeed. About that there is no question. Most likely, their partisan enthusiasm crept into their work even while they were trying to be judicious. After all, Curtiss was particularly famous for tinkering with things until he got them to work.

Conspiracy, however, is another matter. Curtiss, Manly, Walcott, and Zahm were all accomplished and honorable men. When the controversy is inspected closely, there is little reason to doubt their actions or their word. Each of them maintained throughout their lives that they had faithfully tried to reproduce Langley's aerodrome to his original specifications.

Adding to their testimony are recollections Henry Kleckler penned shortly before his death. His unpublished account holds particular historical significance because of his intimate involvement with the details of the reconstruction effort and the fact that, of all those involved, he had perhaps the least personal stake in the outcome. As Kleckler puts it: "We absolutely restored everything the way it was, as near as it was possible to do so."

Such testaments are especially important because, even today, most history books accept Orville's charges of collusion and fraud despite the fact that they prove themselves primarily to be the exaggerated claims of a bitter, aging man.

Of course, the Langley restoration did nothing to take away from Orville and Wilbur Wright's brilliant contribution to the airplane. The Wrights' seminal work to build the first controllable airplane is incisive, original, and clear-headed. It has been closely scrutinized by generations of scholars and justly studied as a model for technological innovation and experimentation.

Then, as now, it is beyond dispute that the Wrights were the first to fly, just as it is beyond dispute that Langley's unmanned model flew years before the Wrights began their work in earnest.

Nor would the reconstruction effort change the outcome of *Wright v. Curtiss*. Larger historical forces would intervene to play a hand in that matter. But Orville is wrong if he would have us see the reconstruction of the Langley aerodrome as some kind of trick in which Curtiss and the Smithsonian team had to make many alterations to get the aerodrome to fly.

Quite the contrary. Even assuming Orville's entire catalog of changes is accurate, the rebuilt aerodrome must be seen as having risen with relatively few major changes that could have enhanced its chance for success. The original motor—even in its diminished state—clearly had enough thrust to launch the aircraft; and the basic tandem-wing design, while not providing the lift of a biplane, clearly proved its viability over Lake Keuka. By any estimation, in other words, Langley must certainly be seen to have come close to creating a working prototype.

Perhaps more pertinent, as a number of aviation historians have rightly noted, with or without changes in the reconstruction process, Langley's original aircraft itself suffered from a number of glaring deficiencies. The most obvious, of course, was the lack of any means to land. One wonders whether Langley was so focused on getting airborne that he hadn't considered how best to land safely with a pilot. Langley is known to have been rigid in his engineering approach, and his aerodrome had a number of other important failings as well. The sharp leading edges of its wings produced unnecessary drag, for instance, and their curvature was not the most aerodynamically effective. From such a critical perspective, Langley's aerodrome must be seen as nothing more than a dead end, despite the demonstration made by Curtiss and the Smithsonian team in 1914. No other inventors, after all, would carry forward Langley's unusual tandem-wing design or his primitive balancing scheme.

And yet, oddly enough, precisely the same complaint can be leveled against the early *Wright Flyer,* with its flexible wings, unstable design, large front elevator, and cumbersome skids requiring a half-ton weight and derrick for takeoff. Only the Wrights' biplane design—lifted directly from Chanute (with his magnanimous permission)—would bear fruit in future airplane models.

One of the most acerbic of the Wright critics, Charles Grey, an aviation buff who knew the early Wright and Curtiss airplanes intimately, makes such an argument. As Grey puts it, the intellectual ferment evident in the early, incomplete designs of Langley and many others, "show how ridiculous is the claim that the Wrights 'invented the airplane.'

"The Wright type biplane with the small leading plane and a method of control which hardly anybody other than the Wrights could manage," Grey argues, "killed most of its pilots and was obsolete and out of production by 1912 when many other designs were flying strongly and developing fast."

By contrast, in spite of the court rulings in favor of the Wright patent, Curtiss's innovations are notable for the way they have so often endured and flourished on their own merits. His airplanes introduced many features—like rigid wings, trailing-edge wing flaps, retractable landing gear, and pontoons—that continue to be time-honored elements of aeronautical design nearly a century later. In this way, Curtiss's accomplishments live on even though his story has been largely forgotten.

ALL BUT THE LEGACY

*From time to time numerous aerial craftsmen have
flourished in the world's eye, only to pass presently
into comparative obscurity, while others too neglected
or too poorly appreciated in their own day
subsequently have risen to high estimation and
permanent honor in the minds of men.*
                                        —ALBERT ZAHM, 1914

At its core, the long, bitter fight
between Glenn Curtiss and the Wright brothers pitted the virtues of
open, shared access to innovation against the driving economic
pressure for monopoly ownership, a debate that resonates through
the years. Having accomplished a tremendous breakthrough in aviation, Wilbur and Orville Wright tried to control the development
of the airplane in its first decade through patents and aggressive
business tactics. Ultimately, their effort would fail.

By contrast, Glenn Hammond Curtiss permitted anyone to use
the principles underlying his inventions—a strategy that enor-

mously benefited the emerging industry. Unlike the Wrights, Curtiss believed his inventions and products should succeed or fail in the marketplace on their own merit. This, ultimately, is the way he would have wanted his career to be judged, and it is how it should be judged: by the lasting, unrivaled success of the aeronautical inventions he created.

Orville Wright continued to vigorously prosecute his lawsuit against Curtiss until 1917 when the U.S. government, responding to the overriding requirements of waging World War I, ordered the nation's two largest airplane companies to settle their differences. The fruitless lawsuit had lasted nearly nine years and proved a costly drain on time, energy, and resources for both sides. With the pressure from the government, a cross-licensing agreement, paying modest royalties to both the Wright and Curtiss companies on the sale of all new American airplanes, was drawn up by Henry Ford's lawyer Benton Crisp.

Years before the settlement, Orville had become a millionaire from the airplane business. But during the entire legal proceedings and for the rest of his life, he would make few further contributions to the fast-maturing aviation field. In fact, most of the Wrights' engineering contributions were obsolete well before the conclusion of *Wright v. Curtiss*.

Meanwhile, freed from litigation in 1917, Curtiss was finally able to take full advantage of his company's superior technology. Even prior to the formal end of the lawsuit, contracts began flooding into his firm, including huge wartime orders from the British government. By the time the United States entered World War I in 1917, the Curtiss Aeroplane Company had become far and away the nation's largest airplane manufacturer. The reason was as simple as it always had been: Curtiss built the best airplanes of his day.

•   •   •

And what of the *America* and Curtiss's promise to build an airplane that could cross the Atlantic Ocean?

In August 1914, the war in Europe interrupted events in Hammondsport just as the *America* drew tantalizingly close to carrying the requisite fuel load. Lieutenant Cyril Porte, *America*'s pilot, was hastily called back into active duty in Britain. The project stalled: there could be no transatlantic flight with Europe at war.

In one of his first acts as wing commander of the British Royal Naval Air Service, Porte arranged to purchase the *America* from Curtiss and ordered more like it to be built for Britain. Within months, Britain deployed the *America* to patrol the English Channel and it sank no fewer than three German U-boats. The handful of "Americas" in the British fleet, in fact, would be the only American-built aircraft to see combat in World War I.

Curtiss was just short of the needed thrust in the *America* when the outbreak of war in Europe prevented his planned transatlantic flight. But at the war's end, *America*'s successor, the *NC-4*—designed by Curtiss in collaboration with the U.S. Navy—made the first-ever airborne transatlantic crossing on May 27, 1919. U.S. Navy commander Albert Read piloted the *NC-4* (NC for Navy-Curtiss). It was one of a fleet of four identical planes that left the coast of Newfoundland heading for Plymouth, England. Facing storms, and high seas, it would be the only one of the four to successfully complete the voyage. The flight, which included a 1,200-mile hop from Newfoundland to the Azores islands, predated Lindbergh's non-stop, solo Atlantic crossing by eight years.

Curtiss was understandably elated to have built the airplane that was the first to accomplish the transatlantic crossing he had promised and envisioned. He often later considered it his most important contribution to aviation. Astonishingly, even with the delay of the war, the flight took place just eleven years after Curtiss had flown his fateful first kilometer in the AEA's *June Bug*. In these few years, the sky was conquered as air travel moved from the risky imaginings of a few visionaries to a full-blown industry. Throughout, Curtiss's state-of-the-art aircraft flew in the vanguard, buffeted by the powerful gusts of change at an extraordinary period in history.

Ultimately, though, Curtiss's success in raising Langley's aerodrome helped undermine his place in history. Orville Wright never forgave Curtiss and dedicated himself to a long, bitter feud with the Smithsonian over the incident that raged for the next quarter of a century.

During that time, Orville refused to donate to the Smithsonian the original *Wright Flyer* that had first flown at Kitty Hawk even though the museum badly wanted it for their aeronautical collection. In retaliation for the Smithsonian's role in the aerodrome episode, Orville had the Kitty Hawk plane shipped to the British Science Museum instead, and there it would stay until after his death.

Orville did finally decide to bequeath the original *Wright Flyer* to the Smithsonian Institution but only after demanding a formal apology for the institution's role in the Langley aerodrome affair thirty-four years earlier. He even got then-secretary Charles G. Abbot to vow never to publish or display any statement that even lends the impression that an aircraft design prior to the *Wright Flyer* of 1903 could have been capable of successful piloted flight.

Abbot held out against some of Orville's demands. He refused, given the lack of hard evidence, to impugn the reputation of his pre-

decessor Walcott and the others involved in the reconstruction of
Langley's aerodrome. After lengthy negotiations, however, Abbot
did agree to publish an apology for the confusion and consternation
the incident caused as well as the list that Orville had cataloged of
alleged modifications that the Curtiss team made to the Langley
aerodrome.

As for Curtiss's company: the development of the *Curtiss JN* airplane—the "Jenny"—would ensure its success. After reviewing the
Jenny's specifications and noting its comparatively reasonable price
tag, Winston Churchill, then First Lord of the British Admiralty,
ordered the British military to "accept everything America can produce to these specifications." Ultimately, the British would order
thousands of them, and the Curtiss firm would manufacture more
than six thousand Jennies by the end of World War I. The airplane
was a two-seater, ideal for training; in fact, the vast majority of the
almost ten thousand American wartime pilots were trained to fly
these planes. The Jenny would formally initiate airmail for the U.S.
Postal Service in 1918 and, after the war, the Jenny would become a
favorite plane of the exhibition fliers in the 1920s known as the barn-
stormers.

With the Hammondsport plant swamped with orders for the
Jenny, Curtiss reorganized and expanded to a facility in Buffalo,
New York, employing as many as twenty thousand workers at its
peak of production. Yet the corporate roster retained many familiar
names: Harry Genung served as company vice president and plant
manager; Henry Kleckler worked as a design engineer; Monroe
Wheeler served as the company's general counsel. Also high in the
hierarchy were Charles Manly, Albert Zahm, and a host of others in

the aviation field with whom Curtiss had worked and who were eager to join his activities.

Much later, in 1929, the two large airplane firms merged to form the Curtiss-Wright Corporation. The companies' combined stock was valued at $220 million immediately following the merger.

By this time, given the speed with which the aviation field was developing, both Orville Wright and Glenn Curtiss had backed away from central positions of authority in their respective organizations. It happened much earlier for Orville, but the fact is, it quickly became clear to both men that they were from a different time. Aviation was moving into an altogether new phase of engineering and production, one well beyond their hands-on kind of technical expertise.

In these years, Curtiss became famous for his largesse with former workers and those who had helped him along the way. He would regularly check to make sure his old workers and supporters had everything they might need. For intimates like Kleckler and Genung, Curtiss built houses in Florida and offered them a long stream of gifts over the years.

Orville's world, on the other hand, closed in ever more tightly around him as the years went by. He lived until 1948, but for many years he rarely left Dayton and hardly even left his Hawthorne Hill mansion, choosing to live the life of a recluse and dedicate himself, as he would put it, "to putting Wilbur's papers in order." Orville never piloted an airplane after 1914 or even flew in one after 1918. Much of his work during the three decades after settling his lawsuit with Curtiss was devoted to fiercely guarding the Wright brothers' legacy.

At the time of the corporate merger, while he had no say in the matter, Orville was reported to have fumed not only over the prospect of joining with Curtiss's company, but especially at the indignity of having Curtiss's name precede his in the new company's title. It was, of course, a simple reflection of the relative size and power of the two firms.

Around this time, many prosperous years after their bitter battle, Curtiss wrote Orville a personal note, suggesting a meeting and a casual talk to lay aside old grudges.

His letter remained unanswered.

A PARTIAL LIST OF INVENTIONS BY GLENN CURTISS[1]

Listed below are some of the 500 inventions credited to Glenn Curtiss, primarily those specifically related to aircraft design. Asterisks denote a device for which Curtiss received a patent.

*Aileron (with Aerial Experiment Association)

Wind wagon for propeller testing

Twist-handle motorcycle throttle control

Wind tunnel design

Shoulder-yoke aileron control

Hydro-aeroplane (now known as seaplane)

*Hydro-aeroplane pontoon

Hydroplane step for pontoons

Tricycle landing gear

Amphibian airplane

Single-hulled flying boat

Machine for forming laminated wood ribs

Laminated-wood propeller and forming machine

---

[1]Adapted from Alden Hatch, *Glenn Curtiss: Pioneer of Naval Aviation* (New York: Julian Messner, Inc., 1942).

Method of joining wood parts in airplane construction

Aerodynamically balanced rudder

Enclosed airplane cockpit

Biplane elevator control system for dirigibles

Steel propeller design

Crankcase reduction gear for propeller drive

Steering system for landing gear

Combined skid and wheel landing gear

Wheel brake for airplanes

Electrically operated throttle control

Retractable landing gear (with Hugh Robinson)

System of compression bracing for wings

Double-surface wings

Watertight double-surface wings

Interplane drift trussing for wings

Pontoon frame construction

*Compartmented pontoon

Propeller-tip reinforcement

Submerged hydroplanes

*Longitudinally continuous pontoon

Friction throttle control

*Dual controls for airplanes

Dual foot control

*Vent tubes to hydroplane step

Ship catapult launching device for aircraft

Aircraft landing and takeoff system from a ship

Wing beam construction

System of airplane anchorage

Folding hood

*Gyroscopic aircraft stabilizer (with Elmer Sperry)

V-bottom flying-boat hull

Multi-engine flying boat

Life preserver design

Adjusting and locking mechanism for retractable landing gear

Detachable airplane wings

Airplane drag brake

Folding operating brace for control surfaces

Tank suspension for upper wing

Streamlined radiator design

Streamlined landing gear

S O U R C E S

George Santayana, the Spanish-American philosopher (and contemporary of Glenn Curtiss's), is famous for noting that those who do not remember the past are condemned to repeat it. But Santayana also proposed that "history is always written wrong. So it always needs to be rewritten." The sentiment propelled me through many partisan accounts of aviation's early history and will, no doubt, encourage others to revise my interpretations and correct my mistakes.

Whenever possible, this work derives from primary sources, such as letters, cables, photographs, and journalistic accounts from the period. Many of these materials reside at two archives: the Glenn Curtiss Museum in Hammondsport, New York, and the Smithsonian National Air and Space Museum's Garber Facility in Suitland, Maryland. Of particular help at the Curtiss Museum were copies of the *Bulletin of the AEA,* transcriptions of private cables sent from Jim Smellie's pharmacy, collected oral histories from some of Curtiss's relatives and coworkers, as well as numerous collections of newspaper clippings about Curtiss and voluminous, uncataloged boxes of postcards, photographs, and memorablia related to his life and work. I also benefited from a rare copy of Curtiss's 1912 autobiography, *The Glenn Curtiss Story,* especially its detailed, first-person account of the 1910 Hudson River flight. In addition, the cav-

ernous museum houses a full-scale replica of the *June Bug* and many original Curtiss artifacts, including his early motorcycles and flying boats.

Particularly helpful primary sources in the Smithsonian's collection
include Charles Walcott's personal photographic record of the aerodrome reconstruction, Alexander Graham Bell's voluminous and wide-
ranging aviation scrapbooks—technically labeled the "Early Aeronautical
Newsclipping (Alexander Graham Bell) Collection—1906-1911"—and
important selections from the correspondence of Octave Chanute. Also
helpful were the Glenn H. Curtiss Collection, the Samuel P. Langley Collection, the Early Aeronautical Patent Collection, and the impressive
William Hammer Collection, including many rare programs, schedules,
pamphlets, and even menus from some of aviation's most important
meets and gatherings.

As useful as these resources were, my book also relies heavily on many
excellent books and articles about Curtiss and early aviation. First among
them is Cecil R. Roseberry's meticulous biography, *Glenn Curtiss: Pioneer of Flight.* Interested readers are referred to Roseberry's far more
thorough treatment of Curtiss's life. Roseberry is to be especially com-
mended for interviewing many of the last remaining individuals with personal reminiscences about Curtiss, something that is now, alas,
impossible to replicate. Other noteworthy volumes on Curtiss include
Alden Hatch's 1942 work, *Glenn Curtiss: Pioneer of Naval Aviation,* and
Clara Studer's 1937 book, *Sky Storming Yankee.*

Published works of special note also include: Octave Chanute's
Progress in Flying Machines (1894), a book that more than any other con-
veys the ferment in aviation research around the turn of the century; the
collected papers of Albert Zahm; Samuel Pierpont Langley's posthumously published account of the aerodrome's development; the
Wright brothers' collected papers; Charles Gibbs-Smith's authoritative
volumes on early aviation history—especially the primary source material

he collected about the 1909 *Grande Semaine d'Aviation* in Rheims, France; Jack Carpenter's quirky but useful volume *Pendulum,* a compendium of excerpted source material about Curtiss, the Wrights, Alexander Graham Bell, Henry Ford, and others of the period; David Baker's encyclopedic *Flight and Flying: A Chronology,* offering dates and capsule descriptions of every important milestone in aviation from 850 B.C. through 1991; and Phil Scott's *The Pioneers of Flight: A Documentary History,* a readable and well-chosen selection of writings by many of the important early figures in aviation history including Cayley, Ader, Mouillard, Bleriot, and many others.

A complete bibliography follows:

Abbot, Charles G. *The 1914 Tests of the Langley "Aerodrome."* Smithsonian Miscellaneous Collections. Vol. 103, No. 8. Washington, D.C.: Smithsonian Institution, October 1942.

———. *Samuel Pierpont Langley.* Smithsonian Miscellaneous Collections. Vol. 92, No. 8. Washington, D.C.: Smithsonian Institution, August 22, 1934.

———. *The Relations Between the Smithsonian Institution and the Wright Brothers.* Washington, D.C.: Smithsonian Institution, 1928.

Bell, Alexander Graham. *The Alexander Graham Bell Family Papers, 1862–1939.* Online collection of the U.S. Library of Congress. Available at: http://memory.loc.gov/ammem/bellhtml/bellhome.html.

———. "Aerial Locomotion," address before the Washington Academy of Sciences, December 13, 1906. *Proceedings of the Washington Academy of Sciences* VIII (March 1907): 407–48.

Brashear, John A. *The Autobiography of a Man Who Loved the Stars.* Ed. W. Lucien Scaife. Boston: Houghton Mifflin Company, 1925.

Brewer, Griffith. "The Langley Machine and the Hammondsport Trials." *Aeronautical Journal* 25, no. 132 (December 1921): 620–44.

Brown, Carrie. "Man in Motion." *Invention and Technology* (Spring 1991): 50–57.

Brown, R. J. "Alexander Graham Bell and the Garfield Assassination." *The History Buff.* Online resource at www.discovery.com.

Bruce, Robert V. "Alexander Graham Bell and the Conquest of Solitude." In *Technology in America: A History of Individuals and Ideas,* ed. Carroll W. Pursell Jr. Cambridge, Mass: MIT Press, 1981.

Callander, Bruce D. "Five Smart Men Who Didn't Invent the Airplane." *Air Force Magazine,* (January 1990): 88–94.

———. "The Critical Twist." *Air Force Magazine* (September 1989): 150–56.

Casey, Louis. *Curtiss: The Hammondsport Era.* New York: Crown Publishers, 1981.

Cayley, George. "On Aerial Navigation." *Nicholson's Journal of Philosophy* XXIV (1809): 164–74. Reprinted in James Means, ed. *Aeronautical Annual,* 1895.

Chaikin, Andrew. *Air and Space: The National Air and Space Museum Story of Flight.* Boston: Bulfinch Press, 1997.

Chanute, Octave. *Progress in Flying Machines.* New York: American Engineer and Railroad Journal, 1894. Reissued Mineola, N.Y.: Dover Publications, 1997.

———. "Scientific Invention." Address before the American Association for the Advancement of Science, August 1886. *Proceedings of the American Association for the Advancement of Science.* Vol. XXXV. Salem, Mass.: Salem Press, 1886.

Crouch, Tom D. *The Bishops Boys: A Life of Wilbur and Orville Wright.* New York: W. W. Norton, 1989.

———. "The Feud Between the Wright Brothers and the Smithsonian." *Invention and Technology* (Spring 1987): 34–46.

———. *A Dream of Wings: Americans and the Airplane, 1875–1905.* New York: W. W. Norton, 1981.

Curtiss Aeroplane Company. *Curtiss Aeroplanes.* Catalog booklet published by the firm c. 1912. Reprinted by the Glenn H. Curtiss Museum, Hammondsport, N.Y., 1986.

Curtiss, Glenn H. "The Commercial Side of Aviation: Business Possibilities of the Aeroplane." *The Saturday Evening Post.* October 1, 1910.

———, and Augustus Post. *The Curtiss Aviation Book.* New York: Frederick A. Stokes, 1912.

Dollfus, Charles, and Henri Bouche. *Histoire de l'Aeronautique.* Paris: Editions Saint-Georges, 1942.

Eklund, Don Dean. *Captain Thomas S. Baldwin: Pioneer American Aeronaut.* Unpublished Ph.D. dissertation, University of Colorado, 1970.

Flink, James J. *America Adopts the Automobile, 1895–1910.* Cambridge, Mass.: MIT Press, 1970.

Gibbs-Smith, Charles H. *The Rebirth of Aviation 1902–1908.* London: Her Majesty's Stationery Office, 1974.

———. *Aviation: An Historical Survey from Its Origins to the End of World War II.* London: Her Majesty's Stationery Office, 1970.

———. *Sir George Cayley, 1773–1857.* London: Her Majesty's Stationery Office, 1968.

———. *The Invention of the Aeroplane (1799–1909).* New York: Taplinger, 1965.

Goldstrom, John. *A Narrative History of Aviation.* New York: Macmillan Company, 1930.

Grahame-White, Claude. *The Story of the Aeroplane.* Boston: Small, Maynard and Company, 1911.

Grey, Charles G. *The Air Cadet's Handbook on How an Aeroplane Flies.* London: G. Allen and Unwin, 1941.

Grosvenor, Edwin S., and Morgan Wesson. *Alexander Graham Bell: The Life and Times of the Man Who Invented the Telephone*. New York: Harry N. Abrams, 1997.

Hammondsport Herald. Selected editions, 1903–1915.

Hatch, Alden. *Glenn Curtiss: Pioneer of Naval Aviation*. New York: Julian Messner, Inc. 1942.

Hodgins, Eric. "Heavier Than Air (Profile of Orville Wright)." *New Yorker* (December 13, 1930): 29–32.

Hodgins, Eric, and F. Alexander Magoun. *Sky High: The Story of Aviation*. Boston: Little, Brown and Company, 1929.

Howard, Fred. *Wilbur and Orville: A Biography of the Wright Brothers*. New York: Ballantine Books, 1987.

Hughes, Thomas P. *American Genesis: A Century of Invention and Technological Enthusiasm 1870–1970*. New York: Viking, 1989.

Jakab, Peter L. *Visions of Flying Machine: The Wright Brothers and the Process of Invention*. Washington, D.C.: Smithsonian Institution Press, 1990.

Josephy, Alvin M. Jr., ed. *The American Heritage History of Flight*. New York: Simon and Schuster, 1962.

Kelly, Fred C. *The Wright Brothers: A Biography Authorized by Orville Wright*. New York: Harcourt, Brace, 1943.

Kelly, Fred C., ed. *Miracle at Kitty Hawk: The Letters of Wilbur and Orville Wright*. New York: Farrar, Straus, and Young, 1951.

Kirk, Stephen. *First in Flight: The Wright Brothers in North Carolina*. Winston-Salem, N.C.: John F. Blair Publisher, 1995.

Langley, Samuel Pierpont, and Charles M. Manly. *Langley Memoir on Mechanical Flight*. Vols. 1 and 2. Washington, D.C.: Smithsonian Institution, 1911.

Larson, George C. "Glenn Curtiss: The Innovator." *Business and Commercial Aviation* (February 1982): 44–46.

Levinson, Nancy Smiler. *Turn of the Century: Our Nation One Hundred Years Ago.* New York: Dutton, 1994.

Loening, Grover. *Our Wings Grow Faster.* New York: Doubleday, Doran and Company, 1935.

Marrero, Frank. *Lincoln Beachey: The Man Who Owned the Sky.* San Francisco: Scottwall Associates, 1997.

McFarland, Marvin W., ed. *The Papers of Wilbur and Orville Wright.* Vol. 1, *1899–1905.* Vol. 2, *1906–1948.* New York: McGraw-Hill, 1953.

Means, James, ed. *Epitome of the Aeronautical Annual.* Boston: W. B. Clarke, 1910.

Mouillard, L. F. *L'Empire de l'Air:* Paris: 1881. Extracted and translated version under title "The Empire of the Air" reprinted in *Ann. Rep. Smithsonian Inst.* (1892): 397–463.

Munson, Kenneth. *Pioneer Aircraft* 1903 14. New York: Macmillan, 1968.

Newcomb, Simon. *Side-lights on Astronomy and Kindred Fields of Popular Science: Essays and Addresses by Simon Newcomb.* New York: Harper and Brothers, 1906.

Park, Edwards. "Langley's Feat and Folly." *Smithsonian* (November 1997) 30 34.

Pisano, Dominick A., and Cathleen Lewis, eds. *Air and Space History: An Annotated Bibliography.* New York: Garland Publishing in association with National Air and Space Museum, 1988.

Post, Augustus. "The Evolution of a Flying Man: Incidents in the Experience of Glenn H. Curtiss with Motors and Aeroplanes." *Century Magazine* (c. 1910).

Prendergast, Curtis. *The First Aviators.* Alexandria, Va.: Time-Life, 1980.

Rendall, Ivan. *Reaching for the Skies.* London: BBC Books, 1988.

Reynolds, Bruce. "George Hallett: Pioneer Mechanic." *Flying* (April 1958): 40–68.

Root, Amos. "What Hath God Wrought?" *Gleanings in Bee Culture* (January 1905). Reprinted in Phil Scott, ed. *Pioneers of Flight,* 135–36.

Roseberry, Cecil R. *Glenn Curtiss: Pioneer of Flight.* Garden City, N.Y.: Doubleday, 1972. Reissued, Syracuse, N.Y.: Syracuse University Press, 1991.

Rotch, A. Lawrence. *The Conquest of the Air.* New York: Moffat, Yard and Company, 1909.

Scharff, Robert, and Walter S. Taylor. *Over Land and Sea.* New York: David McKay, 1968.

Scott, Phil, ed. *The Pioneers of Flight: A Documentary History.* Princeton, N.J.: Princeton University Press, 1999.

———. *The Shoulders of Giants: A History of Human Flight to 1919.* Reading, Mass.: Addison-Wesley, 1995.

———. "Wright v. Curtiss." *Air and Space* (June/July 1997): 66–71.

Sears, Stephen W. "The Intrepid Mr. Curtiss." *American Heritage* (April 1975): 60–65.

Seely, Lyman J. *Flying Pioneers at Hammondsport, N.Y.* Hammondsport Finger Lakes Association and Better Hammondsport Club, 1929.

Smithsonian Institution. *An Account of the Exercises on the Occasion of the Presentation of the Langley Medal, May 6, 1913.* Publication 2233. Washington, D.C.: Smithsonian Institution, 1913.

Stockbridge, Frank Parker. "Glenn Curtiss: Air Pilot No. 1." *Popular Science Monthly* (March 1927) 20–132.

Studer, Clara. *Sky Storming Yankee: The Life of Glenn Curtiss.* New York: Stackpole Sons, 1937.

Veal, C. B. "Manly, The Engineer." Paper presented at the Annual Meeting of the Society of Automotive Engineers, Detroit, Michigan, January 10, 1939. *SAE Transactions,* (April 1939): 145–58.

Winters, Nancy. *Man Flies: The Story of Alberto Santos-Dumont.* Hopewell, N.J.: Ecco Press, 1997.

Wohl, Robert. *A Passion for Wings: Aviation and the Western Imagination, 1908–1918.* New Haven: Yale University Press, 1994.

Worrel, Rodney K. "The Wright Brothers' Pioneer Patent." *American Bar Association Journal* (October 1979): 1512–18.

Wraga, William. *Official History of the Curtiss-Wright Company.* Online at: www.curtisswright.com.

Zahm, Albert F. *Aeronautical Papers 1885–1945.* Notre Dame, Ind.: University of Notre Dame, 1950.

———. "Some Memories of Mr. Curtiss." *National Aeronautic Review* (August 1930).

———. "The First Man-Carrying Aeroplane Capable of Sustained Free Flight: Langley's Success as a Pioneer in Aviation." In *Smithsonian Report 1914,* 217–22. Washington, D.C.: Government Printing Office, 1915.

———. *Aerial Navigation: A Popular Treatise on the Growth of Air Craft and on Aeronautical Meterology.* New York: Appleton, 1911.

ACKNOWLEDGMENTS

My foremost debt of gratitude goes to my family. Books make greater and more insidious demands than other writing projects—and not just upon their authors; my wife, Laura, and children, Elise and Benjamin, supported and sustained me throughout my extended flights into the early twentieth century.

Katinka Matson, my agent, strongly encouraged me to experiment with the narrative form and backed the project generously throughout with her time and energy. I owe special thanks to Terry Karten, my editor at HarperCollins, for her commitment to the book from the outset and her patience in seeing it through. She and her associate, Andrew Proctor, offered many astute and welcome editorial suggestions.

A special note of thanks goes to Marc Miller who—possibly topping his invaluable help with my previous two books—worked his editorial magic on my early drafts. Marc, my longtime friend and former editor, was the first to read these pages and his keen editorial eye and copious corrections proved indispensable.

I also want to acknowledge my extended family for their many kindnesses, especially my father, Roy Shulman, for his interest and unflagging moral support. Heartfelt thanks to: John and Mickey Reed, Sarah Shulman and Tom Garrett, Jill Shulman, Karen Reed, Molly Reed, Christine Reed, Roberta Reed, and Lisa and Peter Crozier. Thanks also to my late grandmother Lena Wolf, who sadly passed away before she had the chance to see this book in print. All these members of my extended family contributed directly or indirectly to making this book possible.

Finally, I owe a special debt to several precious archives and the people who staff them. Most important are Chris Geiselmann and Kirk House at the Glenn Curtiss Museum in Hammondsport, New York. These two made their rich archive available to me in its entirety. They offered their wisdom about surviving Curtiss materials, gave me open access to their copy machine, and later helped with photo reproduction and fact checking. Their museum is a small, out-of-the-way gem for those interested in the early history of aviation, housing everything from complete sets of the *AEA Bulletin,* to a vast cache of memorabilia donated by many Curtiss associates. I thank them both immensely.

By contrast, the staff at the Smithsonian's archive of the National Air and Space Museum at the Garber Center in Suitland, Maryland, brooked no browsing, offered no card catalog, and otherwise put up small roadblocks to research at every turn. Nonetheless, the Smithsonian archive—doubtless the world's single most important aviation repository—was a gold mine of documentary material for this project. In particular, the chance to peruse Alexander Graham Bell's voluminous scrapbooks and Charles Walcott's personal photographic record of the reconstruction of the Langley aerodrome in their original, leather-bound binders did as much to transport me into Glenn Curtiss's story as any other single experience.

I benefited, too, from access to the excellent collections at MIT, Boston College, Stanford University, Cornell University, as well as the Hammondsport Public Library, the Boston Public Library system, and numerous web-based collections on early aviation. It is not an exaggeration to say that these repositories sustain a vital link to our history. Without them—and generous public access to them—books like this one could never be written.

INDEX

Abbot, Charles G., 226–27

Ader, Clement, 98, 144

AEA Bulletin, 133

Aerial Experiment Association
    (AEA), 145
  death of Tom Selfridge and, 146
  formation of, 113, 117–21
  glider *Cygnet* developed by,
    129–30
  June Bug aircraft developed by,
    122–29, 137–43
  move to Hammondsport, NY,
    130–31
  Red Wing biplane developed by,
    131–32
  White Wing aircraft developed by,
    136–37

Aéro Club de France, 150, 165

Aero Club of Albany, 190

Aero Club of America, 35, 72, 94,
    181
  observation of *June Bug* flight by
    members of, 138–43

Scientific American trophy and,
    126–27

Aero Club of New York, official
    observer of Hudson-Fulton Prize
    flight, 189, 191

Aerodrome (aircraft)
  craftsmanship in construction of,
    26
  earlier models of, 7–8, 12
  engine, 10–12, 13, 19, 34–35, 65,
    221
  evaluation of viability of, 221
  first test flight (1903), 1–21
  houseboat and catapult launch for,
    3, 4–5, 6, 9, 14, 18
  C. Manly as pilot of, 5, 6, 11,
    13–14, 16–17
  propeller blades, 65
  reasons for crash of, 17–19
  restoration. *See* Aerodrome restora-
    tion
  tail, 6, 14
  wings, 6, 12

Aerodrome restoration, 25–27,
   33–40, 43, 58–67, 212–22
   conflicts of interest in, 48–49, 58,
      65–66, 219–20
   modifications made in, 35, 48, 49,
      64–65, 217–19, 221
   significance of, 217–22
   successful launch and flight of,
      215–17
   water launch and pontoons on,
      63–65
   O. Wright's opposition to, and
      attempts to discredit, 46–49, 64,
      217–19
Aeronautics (periodical), 58, 173
Ailerons (wing flaps), 45–46, 58
   comparison of, to Wrights'
      method, 160, 177–78
   invention of, 133–34
   on *June Bug* and *White Wing* air-
      craft, 123, 133–36
   lawsuit involving Wright patent
      and, 160, 171, 173, 177–78
   nonsimultaneous, 69
   patent for, granted to Glenn Curtiss
      (1914), 210
Aircraft
   aerodrome. *See* Aerodrome (air-
      craft); Aerodrome restoration
   ailerons. *See* Ailerons (wing
      flaps)
   cockpit, 73
   Cygnet glider, 129–30

dirigibles, 81–94, 100–102
   first amphibious, 188–89
   first commercially sold, 147
   first multiple loops in, 37
   Gold Bug, 53, 147, 159
   June Bug, 122–29, 137–43
   landing gear, retractable, 208
   S. Langley's aerodrome, first flight
      test of (1903), 1–21
   origins of airplane, 94–100
   ornithopter, 155
   pontoons, 63–65, 207
   propeller blades, 65
   Red Wing biplane, 131–32
   seaplanes, 61, 188–89, 207–8
   tail. *See* Tail(s), aircraft
   tandem-wing design for, 6, 12, 40,
      217, 218
   for transatlantic flight (*America*),
      71–78, 212, 225–26
   White Wing, 136–37
   wings. *See* Wing(s), aircraft
   Wright Flyer. See Wright Flyer (air-
      craft)
   Wright patent on stabilization of,
      42, 44–46, 50–51, 135
Aircraft carrier, 208
Airfoil, 137
Airmail, 202, 203, 227
Air Tournament of Los Angeles
   (1910), 176–77
Albany Flier (aircraft), 188–89
   on Hudson River flight, 194–200

Albany-Manhattan flight, G. Curtiss's solo, 186–204
  airplane design and modifi-cations for, 187, 188–89
  Curtiss's financial problems and decision to attempt, 186–87
  finding takeoff and landing sites for, 190–92
  first and second sections of, 195–200
  Hudson-Fulton Prize and, 187, 189–90, 201–3
  planning for, 188–90
  press interest in, 189–90, 193
  significance of, 203–4
  takeoff, 192–95
  third section and landing in Manhattan, 200–201
Allen, James, 38
America (aircraft), 71–78, 212, 225–26
Amphibian plane, G. Curtiss
  modifications and creation of first, 188–89
Army, U.S., 145
Atlantic Ocean, aerial crossing of, 71–78, 212, 225–26
L'Auto (periodical), 170
Automobile, patents and lawsuits regarding, 70, 174–75
Avery, William, 130
Aviation
  early-20th-century technological advances and, 36–37
  fatalities in early, 40, 146
  origins of airplane and history of early, 94–100
  speed record (1909), 162
  Wright brothers' attempt to monopolize, 43–44, 223. See also Patent, Wright brothers'
  Wright brothers' effect on development of, 51, 57–59
Aviation exhibitions and contests
  Dayton, Ohio (1906), 81–85
  Grande Semaine d'Aviation, Rheims, France (1909), 144–66
  Hudson-Fulton Prize, 186–204
  Louisiana Purchase, in St. Louis (1906), 89–90
  Scientific American trophy, 126–17
Aviators, 131
  alienation of, from Wright brothers, 44, 57–59, 160
  Captain T. Baldwin, 81–94, 100–102, 118, 120, 131
  Glenn Curtiss as. See Curtiss, Glenn Hammond
  at Rheims tournament, 154–55, 158
  A. Santos-Dumont, 81–82, 86, 90, 134
  stunt, 36–37, 67
  Wright brothers. See Wright brothers
Awards
  Collier Trophy, 210

Awards (*cont.*)
  Hudson-Fulton Prize, 187,
    189–90, 201–3
  Langley Medal, 38, 210
  Gordon Bennett Trophy, Rheims,
    France (1909), 160–66
  Scientific American trophy, 126–27

Baldwin, F. W. "Casey," Aerial
    Experiment Association and role
    of, 109, 110, 119, 122, 132, 146
Baldwin, Thomas "Captain," career
    and dirigible flights of, 81–94,
    100–102, 118, 120, 131
Balzer, Stephen, 11 n.
Baseball, 67–68
Beach, Stanley Y., 138
Beachey, Lincoln (Link), 36–37, 43,
    67, 212
  dirigible flight (1906), 93
Beard, Luther, 53
Bell, Alexander Graham, 28, 38, 171
  on aerodrome restoration, 63
  early career and inventions of,
    114–17
  glider design by, 129–30
  interest in S. Langley's aerodrome,
    7, 15, 20, 37, 117, 216
  role of G. Curtiss and, Aerial
    Experiment Association,
    103–21, 129–30, 133
  telephone invented by, 43, 44, 94,
    111, 114–15

  wing design suggestions of,
    135–36
Bell, Edward, 116
Bell, Elsie May Gardiner, 110
Bell, Mabel, 110–11, 113, 117, 119
Belmont, August, 44
Benner, Hank, 75
Bennett, James Gordon, 152–53,
    164
Bennitt, Malinda, 30
Biplanes
  June Bug, 122–29, 137–43
  Red Wing, 131–32
Bishop, Cortlandt, 148, 152, 153,
    156, 164, 165, 171
Bleriot, Louis, 194
  flight of, across English Channel
    (1909), 149, 203
  at Rheims tournament (1909),
    154–55, 156, 163–64, 165
Boston American (periodical), 67
Boston Transcript (periodical), 211
Boulton, M. P. W., 134
Brashear, John A., 38, 39
Brewer, Griffith, on aerodrome
    restoration, 47–49, 217, 218–19

Cartier, Louis, 86
Cayley, George, 95–97
Chambers, Washington Irving, 208
Champlin, Harry, 122–23, 139
Chanute, Octave, 89, 120, 124, 130,
    135, 146

aviation information collected and
disseminated by, 54–55, 98–99
criticism of Wrights by, 55–56,
172–73
on S. Langley's aerodrome, 37–38
Churchill, Winston, 227
Clarke, J., 67, 68
Cockburn, George, 162
Cockpit, 73
Collier, Robert, 44
Collier Trophy, 210
Collins, M. P., 199
Cox, James, 47
Crisp, W. Benton, 69–70, 212, 224
Curtiss, Glenn Hammond, 28–31
    Aerial Experiment Association air-
    craft and role of. *See* Aerial
    Experiment Association (AEA)
    aerodrome restoration directed by.
    See Aerodrome restoration
    aileron wing designs of, 45–46,
    123, 160, 177–78
    aircraft developed by. *See Albany
    Flier* (aircraft); *America* (aircraft);
    Curtis JN ("Jenny" aircraft);
    Gold Bug (aircraft); *June Bug*
    (aircraft); *NC-4* (aircraft); *Red
    Wing* (aircraft); *Triad* (aircraft);
    White Wing (aircraft)
    Albany-Manhattan solo flight by, for
    Hudson-Fulton Prize, 186–206
    business relationship with A. Her-
    ring, 146–47, 170–71, 181–82

    business successes of, 206–7,
    227–29
    celebrity of, 164–66, 169–70
    demonstration flights by, 171–72,
    176–77
    engine designs by, 30–31, 82,
    87–89, 92–93
    financial backing for transatlantic
    air crossing proposed by, 71
    financial difficulties of, 181–85,
    186–87
    first air travel by, 92
    first pilot license earned by, 142
    H. Ford's assistance to, 69–70,
    175–76, 211–12
    injunctions against, 41–43, 176,
    186–87, 206
    innovations and inventions of, 29,
    64, 70, 104–5, 137, 207–10,
    222, 223–24, 231–33
    as *June Bug* pilot, 140–43
    on S. Langley, 38–39
    love of experimentation, 208–9
    marriage to Lena Curtiss, 87–88
    meetings and contacts between
    Wright Brothers and, 81–85,
    100–102, 120, 179, 184, 229
    motorcycles built by, 29, 30–31,
    87
    participation in Rheims, France,
    aviation contest, 144–66
    public sentiment in favor of, 69
    son Glenn Jr., 207

Curtiss, Glenn Hammond (*cont.*)
  visit with A. G. Bell and formation
    of Aerial Experiment Associa-
    tion, 103–21
  Wright lawsuits against, 41–46,
    57–59, 68–70, 72–73, 158–60,
    171–87, 205–6, 210–12, 221,
    224
Curtiss, Lena, 108, 118, 140, 144,
    149, 159, 207
  G. Curtiss's Albany-Manhattan
    solo flight and role of, 187, 190,
    193, 194, 195, 197, 201
  first flight of, 189
  marriage to G. Curtiss, 87–88
  son Glenn Jr., 207
Curtiss Aeroplane Company, 25, 61,
    207, 224, 227–29
  employees of, 31–32
Curtiss JN ("Jenny" aircraft), 208,
    227–28
Curtiss Motor Company, 207
Curtiss-Wright Corporation, 228–29
Cygnet (glider), 129–30

De la Croix, Félix du Temple, 99
Dienstbach, Karl, 138
Dirigible, 81–94, 100–102
Doherty, Tony, 41

Ellyson, Theodore "Spuds," 32, 208
Engine(s), 96–97
  for *America*, 73, 76–77

  for T. Baldwin's dirigible, 82, 87,
    89, 92–93
  G. Curtiss designs for, 30–31, 82,
    87–89, 92–93
  for *June Bug*, 124
  for S. Langley's aerodrome, 10–12,
    13, 19, 34–35
  for *Rheims Racer*, 149, 156
English Channel, first flight across
    (1909), 149
Esnault-Pelterie, Robert, 45 n., 134

Fairchild, Daisy Bell, 110, 140–42
Fairchild, David, 110, 140–41
Fallières, Armand, 157
Fisher, Ward, 148, 156
Flights
  Albany to Manhattan, G. Curtiss's
    solo, 186–204
  first officially observed (1908),
    142–43
  first Wright Brothers' (1903), 15,
    50
  transatlantic, 71–78
Flotation devices, 188–89
Ford, Henry, 28, 61, 186
  his assistance and support for G.
    Curtiss, 69–70, 72, 175–76,
    211–12
  Model T built by, 36
  G. Selden patent and, 174–75
French, John, 157
Fulton, Robert, 203

Garfield, James, 115, 116

Gary, George H., 138

Gaynor, William, 202

Genung, Harry, 142, 144, 149, 187, 207, 227, 228
aeronautical work with Glenn Curtiss, 31, 137–38, 147

Genung, Martha, 31, 142, 144, 149, 187

German Wright Company, 159

G. H. Curtiss Manufacturing Company, 87

Gibbs-Smith, Charles H., 134 n.

Gliders
G. Cayley's, 96–97
Cygnet, 129–30

Gold Bug (aircraft), 53, 147, 159

Gordon Bennett Trophy race, Rheims, France event (1909), 160–66

Grande Semaine d'Aviation, Rheims, France (1909), 144–66
G. Curtiss airplane Rheims Racer entered in, 150, 155–56, 160–64
G. Curtiss's journey to attend, 144–49
Gordon Bennett Trophy race in, 160–66
Wright brothers' lawsuit at time of, 158–60

Graphophone, 115

Grey, Charles, 222

Gyroscopic automatic stabilizer, 208

Hallet, George, 74

Hamilton, Charlie, 57

Hammondsport, New York, Curtiss operations in, 25, 27–28, 130

Hammondsport Herald (periodical), 216

Handlebar throttle control, G. Curtiss's motorcycle, 31

Hargrave, Lawrence, 98

Hawley, Allan R., 138

Hazel, John Raymond, Wright lawsuit and rulings of, 173–79, 182, 183, 206, 210

Herring, Augustus, G. Curtiss's business relationship with, 146–47, 170–71

Hewitt, Fred, 16–17

Hill, Thomas A., 173

Hitchcock, Gilbert, 15

Hot air balloons, 82, 165

Houseboat and catapult launch for S. Langley's aerodrome (1903), 3, 4–5, 6, 9, 14, 18

Howard, Fred, 57

Hubbard, William H., 196

Hudson-Fulton Prize, G. Curtiss's solo flight to win, 186–204

Hudson River flight. See Albany-Manhattan flight, G. Curtiss's solo

International Harvester Corporation, 61

Iron lung, invention of, 116
Isham, William B., 199

Johnson, Walter, work on aerodrome
    restoration, 35, 49
Jones, Charles Oliver, 131
Jones, Ernest L., 138
June Bug (aircraft), 122–29, 137–43,
    145
  design and engine of, 123–24,
    137–38
  flight of (1908), 124–25, 138–43
  Scientific American trophy won by
    (1908), 124, 126–27, 142–43

Keller, Helen, 115
Kimball, Wilbur R., 139
Kleckler, Henry, 31–32, 207, 227, 228
  Aerial Experiment Association and
    role of, 122, 125, 140, 147, 149
  aerodrome restoration and work of,
    26, 33–35, 220
  G. Curtiss's Albany-Manhattan
    flight and role of, 188, 190, 181,
    193, 194, 195, 197

Lake, Simon, 138
Lambert, Comte de, 158
Landing gear, retractable, 208
Langley, Samuel Pierpont, 97, 146,
    205. *See also* Aerodrome (air-
    craft); Aerodrome restoration
  departure of, from aeronautics,
    15–16, 20–21

first test flight of aerodrome and,
    1–21
  information provided to Wright
    brothers by, 54
Langley Medal, 38, 210
Latham, Hubert, 154
Lawsuits, Wright brothers', against
    Glenn Curtiss, 41–46, 57–59,
    68–70, 72–73, 158–60, 171–85,
    186–87, 205–6, 210–12, 224
Le Bris, Jean-Marie, 98
Lefebvre, Eugene, 158
Lenormand, Sebastian, 92 n.
Leonardo da Vinci, 94, 172
Lilienthal, Otto, 81, 97–98, 135
Lloyd George, David, 157–58
Lloyds of London, 75–76
Loening, Grover, 44, 47, 179, 206
Longwell, Lewis, 32
Louisiana Purchase Exposition, St.
    Louis (1906), 89–90

McClure's (periodical), 7
McCormick, Samuel B., 38, 61
McCurdy, Douglas, Aerial
    Experiment Association and
    role of, 109, 110, 119, 122, 133,
    146
McEwan, James B., 202
McKinley, William, 8, 66, 174
Macomb, Montgomery M., 18
Manly, Charles, 138, 227
  aerodrome engine design by,
    11–12, 34–35

aerodrome restoration and, 61, 212, 219, 220
as pilot in aerodrome test flight (1903), 5, 6, 13–14, 16–17
Masson, Katherine, 74–75
Maxim, Hiram, 99–100, 146
Military interest in aviation, 2–3, 8–9, 18, 93, 225
Montgomery, John, 97
Motorcycles, Glenn Curtiss designs for, 30–31, 87–89
Motors. *See* Engine(s)
Mouillard, Louis-Pierre, 98, 135, 172
Mozhaisky, Alexandr Fyodorovich, 99
Munn, Charles, 127–28, 203

Nash, Francis, 3, 4–5, 17
NC 1 (aircraft), first transatlantic crossing by, 225
Newcomb, Simon, disbelief in aviation possibilities, 1, 10
New York City, G. Curtiss's flight from Albany to, 186–204
New York City Auto Show, 105–6
New York Evening Mail (periodical), 201
New York Herald (periodical), 66
New York Sun (periodical), 67, 68
New York Times (periodical), 9, 15, 20, 42, 48, 59, 216
G. Curtiss's Albany-Manhattan flight and, 190, 201, 204

New York Tribune (periodical), 68
New York World (periodical), 67
G. Curtiss's Albany-Manhattan flight and, 189–90, 199, 201, 202
Northcliff, Lord, 72
Number 14-bis (aircraft), 134
Number 22 (aircraft), 163

Parachutes, invention of collapsible, 92
Passion for Wings: Aviation and the Western Imagination 1906–1918 (Wohl), 53
Patent Office, U.S., 46
Patent, Glenn Curtiss's, for ailerons, 210
Patent, Wright Brothers', 42, 44–46, 50–51, 135, 159
lawsuits to protect, 41–46, 57–59, 68–70, 72–73, 158–60, 171–85, 186–87, 205–6, 210–12, 221, 224
Paulhan, Louis, 154
Wright court injunction against, 180–81
Pegoud, Adolphe, 37 n.
Penaud, Alphonse, 6
Penaud tail, 6
Pfitzner, Alexander L., 131
Pilcher, Percy, 97
Pilot license, first (1908), 142
Polignac, Grand Marquis de, 150
Pontoons, 63–65, 207

Porte, John Cyril, 74, 75, 225

Post, Augustus, 131, 138, 191–92,
    193, 213

Press

    G. Curtiss's relations with, 49, 67,
        139, 189–90, 193, 201–2

    interest of, in aerodrome restora-
        tion, 66–68

    interest of, in aviation, 139, 143,
        157, 189–90, 193, 201–2

    at S. Langley's aerodrome test,
        1–2, 15

    Wright brothers' relations with,
        52–53, 127–28

Progress in Flying Machines
    (Chanute), 98

Propeller blades, 65

Public interest in aviation, 67,
    150–51, 157, 193, 200–201

Pulitzer, Joseph, 187

    Hudson-Fulton Prize offered by,
        187–204

Randolph, Wallace F., 2

Read, Albert, 225

Red Wing (biplane aircraft) 131–32

Rheims, France, aviation tourna-
    ment. *See Grande Semaine d'Avi-
    ation,* Rheims, France (1909)

Rheims Racer (aircraft), 150, 152, 158

    design, 155–56

    engine, 149, 156

    performance of, in trophy race,
        160–64

Richardson, Holden C., 47

Robinson, Elmer, 194

Robinson, Thomas, 15

Roosevelt, Edith, 93

Roosevelt, Teddy, 89

Root, Amos, 52

Root, Elihu, 2

Roseberry, C. R., 31, 47, 143

Royal Aeronautical Society, 218–19

Santos-Dumont, Alberto, 81–82, 86,
    90, 134

Scientific American, 83, 127

Scientific American aviation trophy,
    June Bug craft as winner of
    (1908), 124, 126–27, 142–43

Seaplanes, 61, 188–89, 207–8

Sea sled, 70

Seely, Lyman, 69, 176, 209

Selden, George, motorcar patent
    and lawsuits involving, 70,
    174–75

Selfridge, Thomas

    Aerial Experiment Association and
        role of, 110, 119, 122, 125, 127,
        129–30, 132, 133

    death of, 145–46

    Wright patent and, 183

Shaw, George Bernard, 114

Shriver, Tod, 147, 149, 156

Smellie, Jim, 30

Smithsonian Institution, 2, 38, 66

    aerodrome restoration and support
        of, 26–27, 219

Langley Aerodynamical Research
  Laboratory at, 39, 49
  O. Wright's feud with, 226–27
Spanish-American War, 8
Sperry, Elmer, Jr., 208
  aerodrome restoration and, 61
Stunt aviators, 36–37, 67
Sullivan, Anne, 115
Swope, Herbert, 67

Taft, William H., 203
Tail(s), aircraft
  aerodrome, 6, 14
  Penaud, 6
Tandem-wing design, 6, 12, 40, 217,
  218
Technological advances of early
  twentieth century, 36
Ten Eyck, Jacob, 190, 194
Tissandier, Paul, 158
Towers, John, 72
Toy, Joe, 67
Triad (aircraft), 208
Trowbridge, J. T., 118

Vanderbilt, Cornelius, 44
Van Tassel, Park A., 91–92
Voisin, Gabriel, 154

Walcott, Charles, 33, 38, 47
  aerodrome restoration and role
  of, 48–49, 61, 63, 66, 212,
  217
Wall Street Air Trust, 44

Wannamaker, Rodman, 71, 72–73
War Department, U.S., support for
  Langley's aerodrome project
  from, 2–3, 8–9, 18
Warner, A. P., 32
Watson, Thomas A., 117
Wheeler, Monroe, 87, 159, 184,
  207, 227
White, Clarence, 194
White, Henry, 164, 165
White Wing (aircraft), 136–37
Wildman, Francis, 63
Williams, J. Newton, 131
Wind tunnel, 137
Wing(s), aircraft
  flaps on. See Ailerons (wing flaps)
  Langley's tandem-wings, 6, 12, 40,
  217, 218
  Wrights' wing-warping design,
  44–46, 135, 172, 178, 205
Wohl, Robert, 53
World War I, 78, 224–25
Wright, Katharine, 46, 47, 56
Wright, Lorin, 49
Wright, Milton, 46
Wright, Orville. See also Wright
  brothers
  demonstration flights of, in Ger-
  many, 159
  his feud with Smithsonian, 226–27
  on S. Langley's work, 8
  opposition to and discrediting of
  aerodrome restoration, 46–49,
  64, 217–19

Wright, Orville (*cont.*)
  personal obsession with G. Cur-
    tiss, 42, 179–80, 205–6, 211,
    228–29
  public flight (1908) piloted by,
    145–46
  rift with supporters and family,
    56–57
  skepticism of, toward aviation
    researchers, 120–21
Wright, Wilbur, 6, 54. *See also*
    Wright brothers
  death of, 42, 210
Wright brothers, 41–59
  aeronautical contributions of, 29
  aircraft of. *See Wright Flyer*
  aviation industry alienation against,
    44, 57–59
  aviation monopoly sought by,
    43–44, 68, 223
  contribution of, to aviation, 220
  first public flight (1908), 145–46
  flight at Kitty Hawk, N.C. (1903),
    15, 50
  lawsuits filed by, 41–46, 57–59,
    68–70, 72–73, 158–60, 171–87,
    206, 210–12, 221, 224
  meetings and contacts between G.
    Curtiss and, 81–85, 100–102,
    120, 179, 184, 229
  myths surrounding, 43
  patent of, 42, 44–46, 50–51, 135
  press challenges to, 127–28

public sentiment against, 69–70,
    181
  refusal to participate in Rheims avi-
    ation tournament, 148
  secrecy and proprietary attitude of,
    50–57, 82–84, 101, 127–28,
    210–11
  wing-warping design of, 44–46,
    135, 172, 178, 205
Wright Company, 44, 46
  appropriation of Curtiss designs,
    210
  merger with Curtiss company,
    228–29
Wright Company Ltd., 47–48
Wright Flyer (aircraft), 53, 101
  aviators' ranking of, 170
  evaluation of viability of, 222
  first public flight of, 145–46
  at Rheims tournament (1909), 158,
    163
  takeoff and sled runners, 128
  O. Wright's feud with Smithsonian
    and donation of, 226–27

Zahm, Albert, 223, 227
  as head of aerodynamics lab,
    39–40, 49
  launch of *America* and, 74, 76
  oversight of aerodrome
    restoration by, 33, 48–49, 60–67,
    212, 214, 215, 217, 218, 219,
    220